W9-BFY-259

722 MILES

THE BUILDING OF THE SUBWAYS AND HOW THEY TRANSFORMED NEW YORK

CLIFTON HOOD

The Johns Hopkins University Press
Baltimore and London

Printed in the United States of America on acid-free paper

Originally published in a hardcover edition by Simon & Schuster, 1993
Johns Hopkins Paperbacks edition, 1995
04 03 02 01 00 99 98 97 5 4 3

The Johns Hopkins University Press
2715 North Charles Street
Baltimore, Maryland 21218-4319
The Johns Hopkins Press Ltd., London

Library of Congress Cataloging-in-Publication Data

Hood, Clifton.
722 miles : the building of the subways and how they transformed
New York / Clifton Hood.—Johns Hopkins paperbacks ed.
p. cm.
Originally published : New York : Simon & Schuster, c1993.
Includes bibliographical references and index.
ISBN 0-8018-5244-7 (pbk.)
1. Subways—New York (N.Y.) 2. Urbanization—New York N.Y.)
I. Title.
[TF847.N5H66 1995]
388.4´28´097471—dc20 95-18961

A catalog record for this book is available from the British Library.

Map Credits
1, 3, & 5: Cartographic Laboratory, University of Wisconsin–Madison.
2 & 4: by Ozzie Grief.

To my mother, Jane D. Hood,
my sister, Nancy Hood,
and in memory of my father,
R. Clifton Hood

CONTENTS

LIST OF MAPS

INTRODUCTION

I wrote this book for two separate audiences: fellow scholars of urban history and readers who are captivated by the history of cities, particularly New York City.

What subject is better suited to popular interest in history than the New York subway? You can touch the subway, see it, hear it, and ride it. In an age such as ours when scholarly research is increasingly abstruse and prevailing literary styles are spare and abstract, the subway is so concrete that at times it seems unrespectable. And the subway itself belongs to the people. On average, more than 2.7 million New Yorkers ride the underground railway every day. The subway figures so prominently in New York City's daily life that it has been represented in movies, paintings, plays, songs, dances, and other forms of popular culture ever since the first line opened in 1904. I think New York City's subway is magical, too. After spending fifteen years researching and writing about the subway, I am more fascinated by it now than the day I began; its attraction has never diminished.

The subway is massive. With 722 miles of tracks, New York City's subway is the longest rapid transit system in the world, larger than the London underground, the Paris metro, the Tokyo subway, and the Seoul subway. If its tracks were laid end to end, the subway would stretch from Manhattan to Chicago. It boasts sixty-four hundred subway cars and 469 stations. It is so immense that it has a 4,250-person police force, the New York City Transit Police Department, which is larger than the entire police departments of such American cities as Boston, Atlanta, St. Louis, Dallas, Denver, and San Francisco.[1]

The subway is essential to New York City's existence. No other American city depends as much on mass transit. In 1989, 46 percent of all New York City workers used the subway to travel to their jobs, a figure far higher than that for any other U.S. city.[2] Without the subway, the offices on Wall Street and in midtown Manhattan would grind to a halt. The symbol of New York may be the skyscraper, which proclaims the city's wealth, power, and size spectacularly. And, compared to the elegant skyscraper, the subway—which remains hidden from view below the surface and is dirty and often overcrowded—is hardly awe-inspiring. Yet, without the subway, the skyscraper would be an empty shell.

The subway integrated New York geographically, overcoming its river barriers and joining four of its five boroughs together. It shaped the development of many city neighborhoods. Without the subway, these residential sections—which are thickly settled, socially diverse, and consist mostly of rental apartments—would more closely resemble the homogeneous, low-density neighborhoods that are characteristic of other U.S. cities.

Built almost entirely between 1900 and 1940, the subway was financed by the City of New York and two private subway companies. It is significant that the city constructed the gigantic rapid transit system with its own money. Until recent decades the subway received no federal assistance at all. It was the product of an era when New York City commanded the resources to solve its most pressing problems. The forty-year period of subway expansion ended in 1940, shortly before the federal government began siphoning off New York City's wealth through taxes that were not returned in the form of grants and subsidies. In view of the federal government's recent disdain for cities, it is unlikely that such a magnificent subway could be built today. According to the New York City Transit Authority, the subway has an estimated replacement cost of $55 billion in current dollars, approximately equivalent to 375 F-22 fighters or twenty-six nuclear-powered aircraft carriers or thirty space shuttles.[3]

Despite its significance, very little has been written about the New York subway. A few unpublished dissertations and monographs have examined one aspect or another, and a few journalistic accounts are available, but there is no general analysis of the subways' development. This book aims to fill that gap.[4]

Three aspects of the subway are examined: first, its political culture, meaning the values, decisions, and institutional relationships that brought about underground rapid transit's rise and fall; second, its impact on New York City, both as a catalyst for urban spatial expansion and as a source of New Yorkers' changing perceptions of the city; and third, its construction and engineering, as one of the largest public works attempted in any city.

Subway Politics

New York's rapid transit system was created as a response to the extraordinary economic and population growth sustained by the city after the 1820s. The city's geography was poorly suited to this explosive development, and a drive for a high-speed, high-capacity subway began in 1888.

This work concentrates on the years from 1888 to 1953, with the time span divided into two periods. In the early period, from 1888 to 1907, the merchants or business interests dominated, while after 1907 professional politicians made the important transit decisions.

On January 31, 1888, Mayor Abram S. Hewitt, a wealthy iron manufacturer and the son-in-law of merchant Peter Cooper, proposed the construction of a state-of-the-art rapid transit railroad. Hewitt's proposition was not the first subway plan contemplated for New York City; a number of blueprints had been drawn up in previous decades, many of them inspired by the opening in 1863 of the world's first underground railway, in London. What distinguished Abram S. Hewitt's proposal was that it established a critical connection between advanced technology and business-government relations. Convinced that a high-speed subway was needed to serve northern Manhattan, Hewitt made the first substantive proposal for government investment in rapid transit.

Hewitt's innovative formula of government ownership and private construction and operation was eventually adopted by the Chamber of Commerce of the State of New York, the city's most powerful business organization. The Chamber secured the passage of the Rapid Transit Act of 1894, which provided for government investment in a subway and gave the Chamber control of rapid transit planning.

In 1900 the merchant-led Rapid Transit Commission contracted with financier August Belmont to build, equip, and operate New York's first subway; it opened in October 1904, operated by Belmont's Interborough Rapid Transit Company (IRT). Together with an extension to Brooklyn that was then being built, the IRT constituted the first of three major stages of construction. It stimulated vast residential expansion in upper Manhattan and the Bronx.

Although the new IRT subway was a great success, August Belmont's Interborough and the Rapid Transit Commission soon clashed over the company's refusal to build additional lines. This conflict heightened in December 1905 when the Interborough acquired its only competitor, Metropolitan Street Railway, and won veto power over new subways.

The Interborough-Metropolitan merger exposed the flaws of the mercantile political system and created pressure for the adoption of new arrangements. Strong press and popular reaction to the merger benefited a group of progressive reformers who wanted to break the Interborough's subway monopoly and build new lines that would disperse immigrants from crowded urban slums to the outskirts of the city. In 1907 the progressives replaced the old Rapid Transit Commission with the more potent Public Service Commission (PSC). Although the dispute between the merchant-commissioners and their progressive critics revolved around the type of regulatory commission that would govern underground rapid transit, it concerned more fundamental questions of political values, practices, and institutional relationships. In general, subway conflicts involved issues of political culture rather than economic structure or government institutions.

A number of key advances were made during the years 1907 to 1953. For instance, the second stage of subway construction, the dual contract system, was approved in March 1913. The dual system integrated the IRT with another corporation, the Brooklyn Rapid Transit Company, or BRT (later renamed the Brooklyn-Manhattan Transit Corporation, or BMT). It more than doubled the total rapid transit mileage and extended beyond the built-up territory to new parts of the Bronx, Queens, and Brooklyn. In addition, the Independent Subway System (IND), the third and last major phase, entered service between 1932 and 1940; unlike the earlier IRT and BMT lines, the IND was publicly owned and operated. The 1930s and 1940s became the golden age for subway passengers. Packed with all classes and ethnic groups, underground rapid transit was a place of excitement and urban color. Inexpensive and relatively free from crime, the subways enabled poor residents to take full advantage of the city, even during the great depression.

But in reality the subways were in trouble. Beginning with the rise of inflation during World War I, a series of financial crises strained the political arrangements between the City of New York and the two subway companies. The increase in costs that followed World War I produced a crisis because the Interborough and Brooklyn Rapid Transit companies were prohibited from raising the nickel fare. As the value of the fare dropped, the IRT and BRT encountered financial trouble, yet their demands for a fare hike met strong resistance from a public with vivid memories of past railway monopolies and poor passenger service.

This general political failure to correct the subways' finances caused a void that was filled by Mayor John F. Hylan (1918–26), an urban populist who was the dominant figure in transit matters in the early 1920s. Hylan acted within the context of a political culture where public ownership and regulation combined with private operation to encourage competition and discord rather than cooperation, where strong popular pressures for a low fare and for high-quality passenger service intensified the friction between government and business, and where ideological resistance to public investment in rapid transit perpetuated the system's fiscal problems. Hylan skillfully manipulated public resentments, blocking regulatory efforts designed to restore the IRT and BRT via financial concessions from the government.

The great depression devastated the Interborough and the Brooklyn-Manhattan. In 1934, Mayor Fiorello H. LaGuardia sought to reorganize the subways. He thought the unification of the privately operated IRT and BMT and the publicly operated IND in a single, municipally run network would save millions of dollars and put the system on a self-sustaining basis, with the nickel fare.

LaGuardia's view of unification was framed by the prevailing political culture. In a decade when the municipality had lost its self-sufficiency and when Franklin D. Roosevelt's national government was taking responsibility for urban functions such as relief and public housing, the mayor proposed bringing a subway that was hemorrhaging money under direct city operation. The idea of federal subsidies for local railway operation was inconceivable in the 1930s, and neither LaGuardia nor any other leader proposed such a drastic step. LaGuardia retained the existing definition of rapid transit as a business enterprise that had to support itself. Due to its development by profit-making corporations, the subway remained saddled with an ideology of business management long after private operation ceased. This business ideology strengthened the resistance to government investment in rapid transit and discouraged contemporaries from formulating a new concept of the subways as an important public service that deserved government subsidy.

In June 1940 the LaGuardia administration succeeded in acquiring the Interborough and Brooklyn-Manhattan properties. Unification of the subways under public control did not solve transit's fiscal problems, however. After World War II the transit system faced renewed competition from the automobile and had to pay higher labor costs. As a result of this financial emergency, the subways incurred operating deficits that undermined the 1940 unification agreement and created pressure for new political arrangements.

The failure to correct subway finances led Paul Windels, a noted Republican lawyer, to campaign for the elimination of direct municipal operation. Windels, who thought of the subway as a business and adamantly opposed public subsidies for it, advocated the establishment of a public authority to restore businesslike management. Popular support for Windels' transit authority proposal grew in the early 1950s as the subway's operating deficits increased, and the New York City Transit Authority was organized in 1953.

My analysis ends in 1953. In my judgment the creation of the Transit Authority set the stage for the subways' physical deterioration and financial collapse in the 1970s. The Transit Authority contributed to the subways' subsequent physical deterioration by enshrining the ideology of business management, insulating transit management from the public, and lessening the accountability of top elected officials for transit decisions.[5]

The Subways' Impact on New York City

Once built, the subways had a tremendous impact on New York City. Due to their high speed, the subways served as a catalyst for the spatial expansion of this large and far-flung city. The 1904 IRT stimulated the growth of vast areas of Manhattan and the Bronx. The 1913 dual contracts spurred residential expansion across a crescent of undeveloped land that extended from the northern Bronx to Queens and Brooklyn. My investigation was of a single neighborhood, Jackson Heights, Queens, which represented one pattern of subway-inspired development. Jackson Heights was selected because it lies along the IRT subway route that runs through the Steinway tunnel, it is located far enough from Manhattan to indicate the subway's effect on the outer boroughs, and it is the product of a highly capitalized and vertically integrated corporation that represented increasingly influential business city planning.

In addition to these spatial patterns, the subways' effect on New Yorkers' changing perceptions of their city is also explored. The subway quickened the pace of urban life and enabled riders to experience a new technological realm below the surface. It allowed New Yorkers to travel across the city as never before. The subway also increased passenger crowding in the cars and aggravated gender, class, and ethnic tensions.

Subway Construction and Engineering

The building of the underground railway required large engineering and work forces and overcame formidable physical barriers. Chapter 3 examines the building of the 1904 IRT subway, the geological obstacles to underground transit construction, the preparation of an engineering plan, and the construction problems that were encountered. The dual system eliminated the East River as a transport barrier and triggered the development of Queens and outer Brooklyn. Chapter 7 details the construction of one subway tunnel, the Steinway tunnel, which went from Forty-second Street in Manhattan to Long Island City, Queens.

My conclusions about New York City's subway are placed in wider context by drawing comparisons with those of other cities wherever possible. These comparisons are useful for understanding government decisions and political ideologies. In view of New York City's enormous size and heavy dependence on rapid transit, contrasts with foreign cities such as London, Paris, and Tokyo are often more valuable, but few historical accounts have been written of post–World War I rapid transit politics in other cities. As the longest subway in the world, however, and as an essential public service in the largest city in the United States, the New York subway is sufficiently important to be worth book-length study in its own right.[6]

PART I

THE MERCHANTS AND THE SUBWAY

PROLOGUE

ABRAM S. HEWITT

On the third day of October 1901, Abram S. Hewitt was a happy man. The former mayor was being honored at a special meeting of the city's most powerful business organization, the Chamber of Commerce of the State of New York. Its President, Morris K. Jesup, presented Hewitt with an ornate gold medal for having been the driving force behind the new subway, the Interborough Rapid Transit Company's line, which was currently under construction; the audience vigorously applauded him.[1]

Abram S. Hewitt, in his late seventies and suffering from jaundice, no longer cut the lordly figure he once did. Standing at the podium, he looked frail and tired, but in spite of his bad health, he still carried himself with an aura of command. True to his bearing, Hewitt belonged to a tiny circle of New York's wealthiest merchants —the Lows, Morgans, Belmonts, Havemeyers, Steinways—who had dominated the city during the nineteenth century. This mercantile elite helped Hewitt create the new underground railroad.

This ceremony itself demonstrated the supremacy of the mercantile elite. Although the subway would be used by millions from every walk of life when it was completed in two or three years, attendance at this meeting was restricted to influential men, such as merchant Charles S. Stewart, financier and philanthropist William

E. Dodge, President James G. Cannon of Chase National Bank, and former Secretary of the Treasury Charles S. Fairchild. It was being held in a small room in the Chamber of Commerce's Nassau Street building, two blocks north of Wall Street, at the very epicenter of American capitalism; it held only about one hundred people, and there was no room for ordinary men.[2]

What made Abram S. Hewitt's presence in this bastion of wealth and privilege particularly compelling was the fact that he was among the last American politicians born in a log cabin. Many nineteenth-century officeholders laid claim to this cherished symbol of democracy during their election campaigns—some truthfully. Yet Hewitt, who had actually experienced grinding rural poverty firsthand, never did and became a self-made aristocrat who maximized the distance between himself and commoners.

Hewitt was born on July 31, 1822, on a corner of a relative's stony-acred farm in Rockland County, New York, where his father had retreated after the bankruptcy of his cabinet-making business. The Hewitt family was so poor that their diet consisted almost entirely of porridge, with very little meat and none of the fancy imported European delicacies that graced the tables of New York City's blue bloods.

Hewitt escaped this misery through social climbing. He won a special scholarship to Columbia College, where he painstakingly cultivated friendships with well-born classmates, including Edward Cooper, the son of the wealthiest man in New York, merchant prince Peter Cooper. Meeting the Cooper family was Hewitt's lucky break, for he not only entered the iron manufacturing business with Edward Cooper—their firm, Cooper, Hewitt and Company, introduced the open-hearth process into the United States in 1868, at its Trenton works—but he also strengthened the family bond by marrying Peter Cooper's only daughter, Sarah Amelia.

By the 1870s Abram S. Hewitt was rich and comfortable, and the owner of a mansion on fashionable Lexington Avenue and a sprawling country house in Ringwood, New Jersey. He could afford to enter politics, and in 1874 he won the first of five terms in Congress.

In 1886, Hewitt ran for mayor of New York in what proved to be the most turbulent election in the city's history. A series of violent strikes that divided cities along class lines broke out across the United States, climaxing with the death of seven people in Chicago's bloody Haymarket massacre. In this tense climate a New York labor party asked radical economist Henry George to be its mayoral candidate. Its selection of George, a flamboyant orator and the author of a famous book about political economy, *Progress and Poverty,* was a brilliant stroke. Denouncing the plutocracy and demanding the abolition of huge fortunes, George rallied entire working-class neighborhoods to his cause. A few days before the election, thirty thousand people braved a driving rainstorm and marched from Union Square to Tompkins Square in support of George.

New York's Democratic party was divided into two rival factions: the ward bosses of Tammany Hall and the wealthy merchants of the County Democracy. Ordinarily, these two factions—the one working-class and Irish Catholic, and the other aristocratic and Protestant—were at each other's throat. But these were not ordinary times, and both sides feared that Henry George would weaken their hold on city politics. Tammany Hall and the County Democracy hurriedly closed ranks behind Abram S. Hewitt as their common standard-bearer. This Democratic strategy worked: Hewitt received 90,500 votes to George's 68,100. A third candidate who was running as a Republican, twenty-seven-year-old state assemblyman Theodore Roosevelt, trailed far behind the leaders.[3]

This shotgun marriage between the machine and the aristocrats began to crumble right after the wedding. Mayor Hewitt loathed Tammany for two reasons. Two decades earlier, Tammany boss William M. Tweed and his henchmen had dominated the city government and had stolen millions of dollars through rigged contracts, sweetheart deals, and kickbacks. As one of the aristocratic reformers who had deposed the Tweed ring in 1872, Hewitt regarded Tammany as a pack of two-bit wardheelers itching to plunder the treasury again. To keep this from happening, the mayor broke a campaign promise and refused to let Tammany share in the municipal patronage.

More fundamentally, Hewitt opposed Tammany because it represented New York's huge immigrant population. As the successor to New York's first Catholic mayor, William R. Grace, Hewitt wanted to reassert Protestant moral authority over this increasingly foreign-born city.[4] Hewitt scorned the newcomers from Ireland, Italy, Russia, and Austria-Hungary as lesser beings who could never achieve the civilized standards of the native stock and who were useful only as a source of unskilled labor for the country's steel plants, textile mills, and coal mines. Hewitt feared that the immigrants' growing political clout would diminish the power of the native ruling class, and he wanted to restrict suffrage so that fewer naturalized immigrants could vote.

Hewitt viewed the mayoralty as a bully pulpit for upholding "American" moral values. Under his orders the police shut down saloons, dance halls, gambling dens, and pool rooms that furnished an outlet for leisure in poor neighborhoods. Hewitt also refused to attend parades that honored Germans, Irish, and other ethnic groups, arguing that these nationalistic exhibitions discouraged immigrants from cutting their ties to the old country and assimilating into American society. A famous incident involved St. Patrick's Day. For years the municipality had flown the green Irish banner over city hall every St. Patrick's Day as a salute to the sons and daughters of the Emerald Isle. But Hewitt, adamantly opposed to lowering the stars and stripes for a foreign banner, ended this custom in the spring of 1888.

This calculated slap provoked a loud and angry uproar from the politically potent Irish-Americans that proved to be Hewitt's undoing. When the mayor ran for another two-year term that November, his opponent trounced him so badly that he retired from electoral politics and never sought office again. Afterward, Hewitt exercised his power from behind the scenes.[5]

Hewitt's brief and tempestuous mayoralty resulted in one important achievement, an ambitious public works program that communicated a bold vision of urban progress. On January 31, 1888, Hewitt sent a message to the Board of Aldermen proposing the construction of a gigantic rapid transit railroad that would extend from city hall in southern Manhattan to what is now the Bronx. This railway would be the most modern urban transit facility in the world, incorporating advanced technology that would make every existing line obsolete. At a time when horse railway cars crept through the city's crowded streets at five or six miles per hour and steam-powered elevated trains traveled at twelve miles per hour, Hewitt wanted his urban railway to make forty or fifty miles per hour.[6]

Hewitt insisted that state-of-the-art rapid transit was the key to New York's retaining its position as the dominant metropolis in North America. The population of the city had quadrupled since 1820, and New York was running out of space to house its residents. According to Hewitt, the existing omnibuses, horse railways, cable lines, and elevateds were too slow and too crowded to solve this problem.

New York boasted the largest urban transport system in the world. There were 94 miles of elevated railways that skirted the surface congestion and ran from one end of Manhattan Island to the other; there were 265 miles of horse-drawn railways and 137 miles of horse omnibuses that plodded along its traffic-clogged avenues and streets; and there was even a mile-long stretch of cable railway that hauled commuters across the Brooklyn Bridge.[7] But as large and comprehensive as it was, this network was nonetheless completely inadequate for the geographically complex city. Only high-speed, high-capacity transport could open for development the unsettled territory that lay at the top of this long, narrow island—in the Upper West Side, Harlem, and Washington Heights.

There was no time to waste. If the middle class and the skilled working class could not find homes and apartments in upper Manhattan, they would look for housing across the water in Brooklyn (still an independent city in 1888) or New Jersey rather than continue living in crowded, uncomfortable lower Manhattan.

New York City simply could not afford to lose so many people or waste so much land. The mayor conceived of rapid transit as a means of broadening the municipal tax base and financing vital public works needed to guarantee the city's supremacy over rival seaports. As the development of the northern districts swelled the tax rolls, the municipality could afford to dredge the ship approaches, build the new wharves and docks, and repave the streets, giving it a vital edge.

This sweeping view of rapid transit's priority had ramifications for the relationship between government and business. At this time the principal instrument for controlling private transit companies was a long-term franchise that defined the route, the rate of fare, and the motive power. Strict laissez-faire limitations on government intervention in the private sphere forbade the municipality from owning or even regulating its railways beyond the terms of these contracts. The franchise system had not seriously impeded the horse omnibuses, horse railways, or the elevated railways, but Mayor Hewitt realized that a rapid transit railway would be another matter because of its great cost.

To pay for his proposal, the mayor proposed a combined business and government development, later known as the Hewitt formula. Under this innovative plan the municipal government would finance and own the line, and a private company would build and manage it.[8]

Hewitt contended that this rapid transit railroad would be as important as the Erie Canal had been when it was completed in 1825. Just as the Erie Canal had enabled New York to become the nation's dominant metropolis and inaugurated a period of unparalleled prosperity, so rapid transit would enable the city to become the seat of the richest commercial empire on earth. As Hewitt told the Board of Aldermen, rapid transit would promote "the future growth of this city in business, wealth and the blessings of civilization" and confirm "its imperial destiny as the greatest city in the world."[9]

A month and a half after Mayor Hewitt sent his message to the aldermen, his urgent cry for better transport was dramatized by a fierce snowstorm, the famous "Blizzard of '88," which immobilized the entire city. For two days that March snow swept out of New Jersey and across the Hudson, propelled by winds that reached fifty miles per hour. Mountains of snow twenty to twenty-five feet high covered street lamps and brownstone stoops, reaching the second floor of some buildings. With gale force winds reducing visibility to less than half a foot, many pedestrians were blown off their feet and had to crawl on their hands and knees. Some did not make it to safety and froze to death in the streets.

The blizzard shut every transit line in the city. Drivers of horse-drawn omnibuses and railway cars abandoned their vehicles in snowdrifts; elevated trains had to stop running when snow blanketed the tracks and choked the steam engines. Even the five-year-old Brooklyn Bridge closed to prevent anyone from dying of exposure or being blown into the East River.

For several days New York resembled an eerie arctic wasteland where the only sources of transportation were sleds, sleighs, skis, and snowshoes. The city stopped working: Commuters could not get home from their jobs; wholesalers could not deliver milk, eggs, butter, or flour, and some grocers ran out of food; coal wagons could not make their runs, and homes went without heat while the mercury dropped to zero.[10]

The blizzard of '88 confirmed Mayor Hewitt's assertion that the modern city was completely dependent on transport. By defining rapid transit as a government responsibility for the first time and by pressing for a technologically advanced railway, Mayor Hewitt sought to guarantee that the life would never go out of New York again.

Unfortunately, Mayor Hewitt's clarion call fell on deaf ears. The response of the Tammany-dominated and Hewitt-hating Board of Aldermen was predictable; it had already overwhelmingly defeated a bill embodying the mayor's idea and was not about to reconsider its decision. More surprising perhaps was the tepid reaction of many businessmen: They opposed the idea of government investment as a violation of laissez-faire precepts. Even real estate developers, who could usually be counted on to push for any scheme that offered the remotest possibility of increasing property values, resisted Hewitt's proposal. Members of the Real Estate Exchange voiced uncharacteristic scruples about the plan's constitutionality and suggested that the government had no more business building a railway than it did building houses.

To these critics the Hewitt plan seemed visionary and impractical. Only a few prominent business organizations, such as the Chamber of Commerce of the State of New York, rallied to Mayor Hewitt's side.[11]

As the members of the Chamber quit applauding and took their seats, Abram S. Hewitt began to speak. The ceremony had moved him deeply. "Mr. President and gentlemen," he said, "I am sure that you and every member of the Chamber will sympathize with my inability to find suitable words to express my profound sense of the honor conferred on me by the presentation of this beautiful gold medal." To Hewitt this tribute represented the highlight of his long, often frustrating political career. "In the course of a long life, devoted largely to the public service, I have been more accustomed to criticism than to commendation." He treasured his gold medal as his "most prized possession" and as "the crowning glory" of his political career, for it symbolized the Chamber's unflagging devotion to the cause that he had advanced thirteen years earlier: rapid transit construction. Picking up the reins, the Chamber had overcome all political obstacles and turned Hewitt's dream into reality.[12]

Hewitt felt certain the underground would be a great success, but he sensed he would not live long enough to see the Interborough Rapid Transit subway finished. He was correct; he died of jaundice a year and a half before the new wonder opened in October 1904.

Looking confidently ahead, he exulted that "no reasonable limits can be assigned to the future growth of this city in prosperity and grandeur."[13] Hewitt felt optimistic about New York's future because he had witnessed its glorious past. During his own lifetime New York had gone from a little seaport clinging to the bottom of Manhattan island to a giant metropolis commanding a vast hinterland and exerting its might around the world. This dramatic transformation was the product of the commercial elite that had built a series of great public works, which had fattened New Yorkers' purses and improved the quality of their lives, among them the Croton Aqueduct water system, Central Park, and the Brooklyn Bridge. But the first and most important of these public works was the Erie Canal, completed seventy-six years earlier when Hewitt was a two-year-old living on that threadbare Rockland County farm.

1

THE GREAT CITY

The Canal

On October 26, 1825, the *Seneca Chief* left Buffalo, New York. Nine days later it reached Manhattan. The arrival of the *Seneca Chief,* the first boat to pass through the Erie Canal, sparked an extraordinary celebration of New York City as a great seaport and inaugurated a period of growth that culminated in the city's commercial dominance over the nation's municipalities.

This celebration took place on water, not land. When the little canal boat anchored off the Battery, at the southern tip of Manhattan, a magnificent flotilla of forty steamers, ceremonial barges, and pilot boats assembled around it. Flags and banners streamed from their masts, and some of their decks were trimmed with evergreens and flowers. Over eight hundred distinguished guests were aboard for the gala event, including the governor of the state, the mayor of the city, and leading merchants. To add to the festivities, ships' bands played military tattoos that could be heard by the crowds lining the shore.

After exchanging salutes with the shore batteries on Manhattan and Governor's islands and with two passing British sloops of war that were flying American flags from their foretop gallant masts as a gesture of respect, the fleet weighed anchor and sailed out of the bay. The gray November skies were clear and dry, the bay was calm.

At the point off Sandy Hook where New York Harbor meets the Atlantic Ocean, the squadron formed a three-mile-wide circle around the *Seneca Chief* and its tenders. The canal boat had hauled a special cargo on its five-hundred-mile journey through the wilderness and over the mountains that separated Buffalo from New York City: two elegant wooden kegs containing water from Lake Erie. As the rolling waves pitched the boats from side to side, Governor DeWitt Clinton took a dipperful from one of the green-and-gilt kegs and poured it into the ocean. He declared the Erie Canal officially open.

After the ceremony ended and the fleet returned to New York, the *Seneca Chief* tied up to a dock to unload its freight. From its hold longshoremen started to remove the cargo—flour, whitefish, butter, potash, maple, and red cedar from western New York, Pennsylvania, Ohio, and Michigan, the first fruits of the rich bounty from the trans-Allegheny west that would help transform New York into a great city. A new era in the history of the city had begun.[1]

The Merchants

Beginning around 1815 enterprising businessmen made several critical trading innovations that tapped the potential of New York Harbor, enormously increasing the stream of imports and exports that passed through the city. New York City became a gigantic marketplace for merchandise to be bought, sold, and redistributed. As middlemen who made themselves indispensable partners in a range of business deals, the merchants pocketed much of the money for themselves and for their city.

These men who wrote their names on the permanent map of New York came from varied backgrounds. There were descendants of the original Dutch settlers, such as William Bayard, a partner in the mercantile firm of LeRoy, Bayard and Company, who engaged in privateering during the War of 1812, and Cornelius Vanderbilt, who in 1810 began a ferry service from his native Staten Island to Manhattan that became the foundation of the greatest steamship and railroad empire in the United States. There were wealthy old New York families such as the Barclays, Delanceys, and Clintons, who traced their ancestry back to the Knickerbocker aristocracy. There were immigrants such as Archibald Gracie, a Scotsman who built the country house overlooking the East River that today is the official residence of the mayor, and John Jacob Astor, a German immigrant from the town of Waldorf who made a fortune as a fur trader in the Pacific Northwest. There were New Englanders such as the Low family of Salem, Massachusetts, who trafficked in opium in China and India, and Rowland H. Macy, a Nantucket whaling captain who founded a small shop in 1858 that ultimately grew to be a giant among department stores. (According to legend, a red star tattooed on Captain Macy's arm was the source of the company's logo).[2]

New York's elite encompassed a wide range of regional and national differences. This does not mean that upper-class society was an oasis of harmony and brotherhood. Far from it. These blue bloods were certainly not free of the social intolerance that bedeviled the entire country, and outsiders who could conform to the prevailing standards of white Protestant society had a far better chance of being admitted to the upper crust than those who could not.

But the upper stratum of New York was far more diverse than any of its main commercial rivals. Unlike Boston or Philadelphia, New York City had been founded as a trading post rather than as a religious utopia, and commercial values remained predominant as the city grew. Consequently, New York welcomed aggressive outsiders who wanted to make money far more warmly than conservative Boston, Philadelphia, Baltimore, or Charleston, and absorbed many of the newcomers into its highest circles.

In the end, money was what always mattered in New York. Almost any white family that made enough money could eventually gain entry to the upper reaches after passing through the right schools, clubs, and churches and after paying homage to social customs.[3]

Human initiative was primarily responsible for the success of New York City. In competing with other cities, however, New York merchants had the incalculable advantage of being based in the greatest harbor in North America. "The harbour of New York," a local guidebook boasted in 1833, "is safe and commodious, its circumference being about 25 miles, and the largest vessels may come up to the wharves of the city."[4] No other seaport could claim the combination of the short, seventeen-mile-long sea approaches from the bar at Sandy Hook to the berths on Manhattan Island, the roomy interior anchorage that provided sanctuary from the winter storms that battered the coast, and the interior trade route up the Hudson River that put New York City in a class of its own—not Philadelphia, Boston, or Baltimore, not Montreal, Charleston, or New Orleans. These natural assets gave the city's traders a tremendous edge in their battle for commercial supremacy.

Improving on nature's gifts, New York's merchants made three key innovations that helped transform the city. The first involved transatlantic shipping. In 1818 several Quaker merchants, including Isaac Wright and his son William from Long Island, and Francis Thompson and his nephew Jeremiah from Yorkshire, England, inaugurated regular packet service from New York to Liverpool. Before that time ships waited to set sail until their holds were full, forcing passengers and traders to cool their heels before finally casting off. But the Wrights and Thompsons founded a new firm, the Black Ball line, which had four small ships that followed a set schedule of arrivals and departures whether their freight manifests were complete or not. By changing Atlantic shipping from an erratic endeavor to a predictable business, the Black Ballers (so-called because of the distinctive symbol on their foretop sails) stimulated a large increase in traffic. As more and more cabin passengers, fine textiles, specie, newspapers, and letters shuttled across the ocean, New York became the main transfer point for merchandise and information going between North America and Europe. New and more colorfully named packet lines, such as Red Star and Blue Swallowtail, soon competed with the pioneer, but New York was already established as the U.S. terminus, and the upstarts used it, too.

A second advance came when New Yorkers created a triangular cotton route. By the 1820s no commerce was more lucrative than the exchange of raw cotton grown by African-American slaves in Virginia, the Carolinas, and Georgia for the fine clothing, furniture, and other products that French and English workers manufactured; it completely dwarfed New York's flour and meat business as well as its packet traffic. Although the raw cotton could have gone directly from the South to the European ports of Liverpool and Havre where it was processed, New Yorkers wanted a piece of the action. Using their control of shipping companies and bank credit as leverage over southern planters, Jeremiah Thompson, a founder of Black Ball, and iron importer Anson G. Phelps redirected this commerce hundreds of miles off its normal path, sending it through New York Harbor. Under this arrangement, coastal schooners of Union Line, Old Line, Todd's Line, and other firms brought loads of the burlap-wrapped, four hundred-pound bales of cotton to New York for transshipment to ocean packets bound for Europe. On the return voyage the packets carried immigrants and general freight to New York, including consignments that paid for the cotton and were earmarked for Mobile, Savannah, and New Orleans. New York muscled into the richest antebellum commerce of all.[5]

Third, the city's merchants were instrumental in constructing the Erie Canal. The leader of this daunting project was DeWitt Clinton, the scion of an eminent colonial family whose father had fought as a major general in the Revolutionary War and whose uncle had been the first governor of New York State, in 1777. DeWitt Clinton inherited both wealth and political clout from his forebears. An overbearing, cynical man who looked on New York as his personal duchy, Clinton vied with men such as Aaron Burr, Robert R. Livingston, and Martin Van Buren for command of the Republican party and reigned as one of the state's most powerful figures for over a quarter century before his death in 1828. He held almost every government office in New York at one time or another, including mayor of the city for ten years and governor of the state for nine.

Clinton began to press for the construction of a canal across New York State as early as 1810. He argued that New York needed an interior route from the Hudson Valley to the Great Lakes in order to exploit the growing western markets and also to protect against the possibility of a British invasion from Canada. To drum up support for the canal, Clinton lobbied for federal money in Washington, D.C., held dozens of public meetings in towns and villages across the state, and published a series of broadsides that were printed in the popular press. At times Clinton's enthusiasm for the plan bordered on megalomania, and his critics took great delight in ridiculing the idea as "DeWitt Clinton's ditch" and in warning that this costly public work would drive the state into bankruptcy. But Clinton persevered, and in April 1817 the state legislature finally approved the canal. The groundbreaking ceremony took place at Fort Stanwix on the Mohawk River on the Fourth of July 1817, three days after Clinton entered the governor's office.

As Clinton had prophesied, the Erie Canal substantially expanded New York City's hinterland. By cutting freight charges between Buffalo and New York City from $100 to $10 per ton and by reducing hauling time from twenty-six to six days, the waterway made the city a transfer point for grain, lumber, meat, and other products from the Great Lakes region and the upper Mississippi Valley. Indeed, the Erie Canal carried so much traffic that in only seven years the State of New York recovered the $8 million cost of construction.[6]

This canal was the right improvement at the right time. It enabled New York to exploit the water-level route through the Hudson and Mohawk valleys at a time when its rivals remained trapped behind higher mountains. Although the success of the Erie Canal prompted Baltimore, Philadelphia, and Washington, D.C., to try building their own waterways in a desperate bid to keep pace with New York, none of these cities could match the pioneer. Baltimore eventually abandoned its project as a total failure, while massive investments were required to finish Philadelphia's Pennsylvania Portage and Canal system (under construction from 1826 to 1840) and Washington's Chesapeake and Ohio Canal (1828 to 1850). Moreover, both waterways immediately faced strong competition from a new source of transportation, the railroads, which made them obsolete.

The Erie Canal, too, slowly faded in importance, but it made an extraordinary impact on the city long before the New York Central Railroad surpassed it after the Civil War. Together with the triangular cotton trade and regular packet service, it made New York the dominant American seaport. By 1860, New York accounted for one-third of the country's total imports and two-thirds of its exports.

According to historian Robert G. Albion, New York's textile imports were greater than the combined total imports of Boston, Philadelphia, and Baltimore. This supremacy extended to human commerce, too. Of the 5,458,000 immigrants who came to the United States between 1820 and 1860, 70 percent entered the country through New York. By the beginning of the Civil War, New York ranked as the busiest port in the Western Hemisphere.[7]

Standing on the decks of the *Seneca Chief* that November day in 1825 and gazing at the magnificent harbor, DeWitt Clinton predicted that New York would become a "great emporium" with "no limit to your lucrative extension of trade and commerce."[8] Within two generations his vision became reality.

The Urban Archipelago

New York City grew at a remarkable rate during the Erie Canal years. In 1800, New York was a relatively insignificant place. When London and Tokyo had populations of over one million and Paris had approximately 700,000 inhabitants, New York had just 79,216 residents and ranked behind Philadelphia as the second largest city in the United States. Then New York's port began to boom. Hundreds of thousands from northern Europe, New England, and the Middle Atlantic region came to fill the new jobs on the city's docks and, later, in its manufacturing industries. Year after year, decade after decade, the population of New York City soared upward at an amazing pace: It doubled from 1820 to 1840, doubled again from 1840 to 1860, doubled yet again from 1860 to 1890. By 1900 this gigantic metropolis housed a total of 3,437,202 people and was the second largest city in the entire world.[9]

But the geography of New York City was poorly suited to this kind of explosive development. New York is an urban archipelago that sprawls across several big islands. It is divided by major waterways—the upper bay, the Hudson River, and the East River—that present formidable obstacles to travelers.

The center of this urban archipelago is Manhattan Island. Thirteen miles long and two miles across at its widest point, Manhattan contains only twenty-three square miles.[10] For many years the boundaries of New York City were coterminous with Manhattan Island. Not until 1874 did New York City expand across the narrow Harlem River by annexing a portion of Westchester County that lay west of the Bronx River (known then as the annexed district and now part of the Bronx); not until 1898 did the five separate boroughs (Manhattan, Brooklyn, the Bronx, Queens, and Staten Island) unite to form New York City as we know it today.

This combination of arduous island geography and rapid population growth made transportation New York's highest priority during the nineteenth century. Ironically, after spending its first two centuries trying to capitalize on its splendid system of bays and rivers, New York succeeded so well and grew so fast that those waterways became the city's most serious transportation impediment.

The Crowded Island

When the Erie Canal opened in 1825, most of New York's 166,086 residents lived at the southern end of Manhattan Island. The city was thickly settled and highly compact: The built-up zone ended only three miles above the Battery, at Canal Street. New Yorkers could easily walk from one side of town to the other in thirty minutes or less and usually got around the city on foot or horseback. They had no use for passenger vehicles except for the stages, schooners, and steamers that traveled intercity routes.

In contrast, the northern part of Manhattan Island was isolated and sparsely populated. Only one thoroughfare, a narrow, unpaved country lane known as Post Road, ran the entire distance from the Battery to the top of Manhattan; Post Road meandered up the center of the island, ran through the village of Harlem, and then crossed Harlem River via ferry at Kingsbridge. Above Greenwich Village the countryside seemed far removed from the booming seaport and more attuned to the passage of the seasons than to commercial rhythms.

Northern Manhattan contained extremely rugged terrain. The ground was hilly and marked with rocky outcroppings, potholes, and valleys. It was covered with dense patches of wildflowers, shrubs, vines, and trees. Thick stands of hickories, red oaks, tulips, and maples were plentiful. The island also had an abundance of spring-fed ponds and brooks—Sunfish Pond, De Voor's Millstream, Harlem Creek, and Sherman's Creek—where deer, cottontail rabbits, and squirrels appeared to drink and forage for food. Game animals also flourished along the island's shores. Manhattan was almost completely ringed with salt marshes, where brackish water and tall, flowering cordgrass supported a rich variety of wildlife, including egrets, herons, kingfishers, ducks, blue crabs, turtles, and striped bass.[11]

But during the nineteenth century, transportation improvements would erase this bucolic landscape and accelerate the urban development of Manhattan Island.

The Horse Railway

On November 26, 1832, the New York and Harlem Railroad inaugurated the first horse railway in the world, on a city route up the Bowery from Prince Street to Fourteenth Street. The maiden car, named the *John Mason* for the president of the railroad, was assembled by John Stephenson, a twenty-three-year-old native of County Armagh in northern Ireland whose family had emigrated to the United States when he was a baby. Stephenson was trained for the merchant's countinghouse. At sixteen he quit school and took a job as a clerk, but the young man hated the dull, plodding routine of office work. A talented handyman who was always tinkering with wood and machinery, Stephenson desperately wanted to give up shopkeeping and develop his mechanical skills. Three years later he finally persuaded his parents to apprentice him to a master coachmaker, Andrew Wade. He flourished there.

At the end of his apprenticeship, Stephenson opened his own shop in the rear of a stable and began turning out his own vehicles. This marked the beginning of a long and successful career as one of the country's leading tram builders. Before its founder's death in 1893, the John Stephenson Company built cars for railways throughout the United States and in every part of the world, including

France, Argentina, Mexico, New Zealand, South Africa, and India. It produced more than twenty-five thousand cars.

But Stephenson was proudest of the *John Mason*. Manufactured at his Elizabeth Street workshop, the car was a low-slung wooden box that sat on four wheels. It was pulled by a team of two horses and ran on iron rails that were sunk below the surface of the street. Passengers climbed up to the roof and rode behind the driver or else entered through a rear door and sat on seats that ran the length of the cloth-covered interior compartment.[12]

Actually, the *John Mason* was not the first mass transit conveyance to appear on New York streets. That distinction belongs to a type of vehicle called an omnibus that had originated in the French city of Nantes. Learning of this French innovation in the late 1820s, Abram Brower, a local stage coach operator and stable owner, put several omnibuses into service on Broadway. Physically, Brower's omnibuses closely resembled the stagecoaches that for years had been making long-distance runs to Boston and Philadelphia and shorter jaunts to Yorkville, Harlem, and Manhattanville. Like the stagecoach, the omnibus was a simple horse-drawn wagon carrying about twelve to fifteen passengers. What distinguished the omnibus from the stagecoach was not the design but rather the quality of service: Brower's fleet operated along fixed routes, with regular stops, on frequent schedules, and with a set fare. Because passengers had a source of transportation that could be counted on, the omnibus lines registered impressive gains in patronage and ordered new vehicles, many from John Stephenson's company. In the 1830s so many of these gaily decorated vehicles—painted bright orange, yellow, or white, and often emblazoned with coats of arms, landscape scenes, and human portraits—appeared on the avenues that New York was dubbed the "city of omnibuses."[13]

But the horse car was technologically superior to the omnibus. Instead of having to run directly on the rough, poorly maintained New York streets, horse cars rolled over smooth iron rails that reduced friction to a minimum. This improvement had enormous consequences: The horse car made speeds up to eight miles per hour, about one-third faster than the omnibus; it hauled three times more passengers for the same investment of energy; and it permitted a much more comfortable ride. "You glide along very smoothly and very swiftly," English traveler George Augustus Sala reported in 1865.

> You know that although the poor animals are drawing a caravan containing from thirty to forty passengers, the rail makes their task comparatively lightsome, and they are not half so cruelly worked as the omnibus horses on Broadway.[14]

The omnibus had little impact on the expansion of the built-up area, but the faster horse railway triggered a massive construction boom in the outlying sections. By allowing residents to live farther from their places of work without having to spend much more time commuting, the horse railway drastically reshaped the city. Where woods, orchards, and cultivated fields had once stood, buildings suddenly appeared. Between 1832 and 1860 the northern boundary of the zone of concentrated settlement moved from Houston Street to Forty-second Street. This was astonishing: In that brief thirty-year period the urban frontier advanced twice as far as it had in the previous two hundred years. "Overturn, overturn, overturn," exclaimed Philip Hone, a former Federalist mayor and a diarist, "is the maxim of New York."[15]

Upper-class families dominated the new neighborhoods. As warehouses, dry goods stores, and other businesses encroached upon their old homes, wealthy citizens built spacious brownstone row houses on the straight and broad avenues and on the secluded side streets of the new areas. The shady, tree-lined refuges of Great Jones Street, Lafayette Place, St. Mark's Place, and Bond Street became the addresses of some of the nation's most celebrated citizens. Former mayor Philip Hone and merchant Peter Schermerhorn lived on Great Jones Street, while Bond Street was home to General Winfield Scott, the hero of the Mexican-American War, and Albert Gallatin, a founder of New York University and Secretary of the Treasury under presidents Thomas Jefferson and James Madison.[16]

Commuting was out of the question for working-class New Yorkers, however. Take, for instance, the poor Irish immigrants and African-Americans who lived in the notorious Five Points district. Located several blocks north of city hall, centered around the confluence of Baxter, Worth, and Park streets, and built partly on the landfill of the former Collect Pond, Five Points was the worst slum in the Western Hemisphere. Its narrow, crooked streets had a forbidding reputation as dangerous, crime-ridden warrens. The density of population was exceptionally high. The residents were desperately poor people who could not afford decent housing; they lived in cellars, attics, stable lofts, and anywhere else space could be found.

Immigrant families lacked the control over their own lives required to escape this wretched neighborhood. The main problem was the onerous labor market. Jobs were temporary and low-paying, and entire families had to work to make ends meet. To be sure of regular employment, poor laborers often had to live at a central location like the Points, which was within walking distance of the docks, clothing factories, and other labor markets. Working ten-hour shifts and earning $2.00 a day, these men and women could spare neither the time nor the money to move away from the city center. Few immigrants rode the horse cars or saw the grand new neighborhoods on the fringe.[17]

"Modern Martyrdom"

As the horse railway pushed the urban frontier northward, more and more middle-class and upper-class New Yorkers used public conveyances to travel between home and workplace. By 1860, Manhattan's fourteen-horse railway companies carried more than 38 million riders annually, while its twenty-nine omnibus lines operated 671 vehicles, each of which made an average of ten trips uptown and ten trips downtown every business day.[18] An important result of this new prominence was that popular disenchantment with transit escalated. There were essentially two sources of complaint: vehicular congestion on the streets and overcrowding in the cars.

For the first time in New York, traffic emerged as a major problem. At peak periods the streets were a solid mass of braying, animal-powered vehicles. Walt Whitman compared this chaos to that of military battles, where regiments and platoons clashed in violent disarray. Wagons, lorries, carriages, and omnibuses moved at different speeds, maneuvered in and out of traffic, and dodged from one side of the street to another. The twentieth-century pattern of separate streams of vehicles flowing smoothly in opposite directions was largely unknown. An 1866 guidebook described traffic-choked Broadway as "a Babel scene of confusion." Indeed, pedestrians risked their lives trying to cross this busy eighty-foot-wide thoroughfare.[19] In 1837, Asa Greene provided detailed instructions for anyone who dared walk across Broadway:

> To perform the feat with any degree of safety, you must button your coat tight about you, see that your shoes are secure at the heels, settle your hat firmly on your head, look up street and down street, at the self-same moment, to see what carts and carriages are upon you, and then run for your life.

Even so, Greene reckoned that pedestrians would be "exceedingly fortunate if they get over with sound bones and a whole skin."[20]

Conditions inside the cars were equally grim. On October 2, 1864, the *New York Herald* issued a manifesto on the straphanger's plight:

> Modern martyrdom may be succinctly defined as riding in a New York omnibus. The discomforts, inconveniences, and annoyances of a trip in one of these vehicles are almost intolerable. From the beginning to the end of the journey a constant quarrel is progressing. The driver quarrels with the passengers and the passengers quarrel with the driver. There are quarrels about getting out and quarrels about getting in. There are quarrels about change and quarrels about the ticket swindle.[21]

In the words of the newspaper, the omnibus was a "perfect Bedlam on wheels."[22]

The horse railway was not much better. Patrons did enjoy a much smoother ride on the horse cars than on the omnibuses. Because the horse cars imposed a five-cent fare, however, they were much more crowded than the omnibuses, which charged a dime. The *Herald* observed that

> people are packed into them like sardines in a box, with perspiration for oil. The seats being more than filled, the passengers are placed in rows down the middle, where they hang on by the straps, like smoked hams in a corner grocery.[23]

During rush hours the poorly ventilated horse cars had atmospheres of "foul, close, heated air [that could be] poisonous." And when heavy traffic snarled the cars, residents who commuted between downtown jobs and uptown homes sometimes wound up spending an hour and a half in this dreadful environment. The only people who did not object to this inhumane and unsanitary treatment were the pickpockets, who exploited rush-hour overcrowding to lift watches, handbags, and breast pins from weary, unsuspecting riders.[24]

By the 1860s the phenomenal spatial expansion of New York City had clearly outstripped the omnibuses and railways. "Something more is needed," the *Herald* asserted, "to supply the popular and increasing demand for city conveyances."[25]

One solution to this misery was rapid transit, which is defined as any type of mass transit that has its own right of way and does not compete for space with other vehicles; the most familiar examples are elevated and subterranean railways. By skirting the street congestion that delayed omnibuses and horse cars, these trains could achieve much faster speeds, spurring more urban expansion.

After the Civil War, engineers increasingly began to explore the possibilities of rapid transit. One extraordinary man, convinced that the answer lay in going underground in wind-blown trains, concocted a bizarre scheme to build a subway in total secrecy below the busiest thoroughfare in the city.

Alfred E. Beach

In 1868, Alfred E. Beach received permission from the state legislature to construct a pair of pneumatic dispatch tubes underneath Broadway, ostensibly for the transmittal of letters and packages. Actually, the forty-two-year-old inventor and journalist intended to build a passenger subway. The reason Beach kept his real plans to himself and submitted a false and flagrantly illegal prospectus was his fear that corrupt state legislators who were in the pocket of the owners of the city's horse railways would reject his proposal in order to prevent potentially ruinous competition. Beach had good reason for concern: Three years earlier, Michigan railroader Hugh B. Willson advanced a pioneering proposal for an underground railway from the Battery to Central Park, only to run afoul of Tammany boss William M. Tweed, the city's most powerful politician and a large horse railway investor. To no one's surprise, the legislators obediently shelved Willson's bill.[26]

Alfred E. Beach, the scion of a learned Massachusetts clan that proudly traced its lineage to Pilgrim elder William Brewster and to Elihu Yale, the father of Yale College, was born in Springfield, Massachusetts, in 1826. Beach's father, Moses Yale Beach, moved his family to booming New York City, where he founded the *New York Sun*. The young boy grew up in the newspaper's offices; when he was not writing copy or hawking extras on street corners, he was setting type, running the steam press, or balancing the accounts.

At twenty Beach borrowed money from a schoolmate and bought a small technical journal that had been created a year earlier, in 1845. The name of this obscure and struggling publication was *Scientific American*. Under Beach's skilled guidance it blossomed. As the editor and co-owner of *Scientific American* for half a century until his death on New Year's Day 1896, Beach transformed it into perhaps the single most important disseminator of technical and scientific information in North America and a major force in the industrialization of the United States. A bright, dynamic man who combined the best of the New England literary tradition with New York's delight for money-making and its hunger for new ideas, Alfred E. Beach occupies a key place in the history of American technology.

But Beach did not just sit at his desk and scribble about someone else's inventions. A born tinkerer who had the "smart" hands of the mechanically adept, Beach was constantly tearing apart some contraption or other and trying to make it work better. He toyed with telegraphs, telephones, and cable railways, and a typewriter that he built for the use of the blind captured a prestigious gold medal at the Crystal Palace exposition in New York in 1856.[27]

In the mid-1860s Beach enlisted in the crusade to find a new technology to replace the horse as the primary motive power for urban mass transit. One possibility was offered by steam engines, which had compiled a record of safe and reliable operation on boats and long-distance trains. Steam locomotives were employed on the world's first subway, a four-mile segment of London's Metropolitan Railway, from Farringdon Street to Bishop's Road, which opened on January 10, 1863. But many critics believed that steam was too dirty and too noisy for permanent urban use, especially in the confined space of underground tunnels where the filthy exhaust could not be properly vented. (Londoners complained about their smoke-filled cars and stations until the original stretch of the underground finally converted to cleaner electricity in 1905).[28] Accordingly, inventors experimented with a vast array of possibilities, including storage batteries and gravity engines.

In the 1860s the most promising advances were being made by T. W. Rammell, an English engineer who was working with a new source of propulsion known as pneumatic power, or compressed air. In 1861, Rammel installed a quarter-mile-long test track at Battersea Fields outside London. It used a thirty-horsepower steam engine to produce compressed air, which blew a five-foot-long canister through an airtight tube. The success of this model won Rammel a fat contract from the British post office to build an immense network of forty-eight pneumatic tubes crisscrossing London in order to speed the distribution of mail in the world's largest metropolis. It also attracted attention across the ocean in the offices of *Scientific American* at 37 Park Row, New York City.[29]

In 1867, Alfred E. Beach unveiled his own pneumatic system at the American Institute Fair, which took place in Armory Hall on Fourteenth Street in lower Manhattan. Beach exhibited two tubes at the fair. One, a small line designed to carry letters and packages, excited little public comment at the fair, but the second apparatus caused an uproar. Suspended by iron loops from the ceiling and running through midair to every corner of the armory, this six-foot-wide wooden vacuum tube was a passenger subway that became the rage of the city. By the time the fair closed in November 1867, more than seventy-five thousand people had ridden it.[30]

This impressive showing paid large dividends. The following year Beach received approval from the state legislature to build a pair of small pneumatic tubes designed to expedite the distribution of letters and parcels from the post office. But Beach had no intention of settling for such small potatoes. Perhaps the legislators who thought they were authorizing a postal dispatch network should have pondered a pamphlet that Beach had written for the American Institute Fair. In it Beach envisioned the construction of a system of pneumatic tubes that would be capable of transporting people from city hall to Madison Square in five minutes, to Central Park in eight minutes, and to Washington Heights in twenty minutes. Beach's plan was simple if wildly unorthodox. Convinced that the graft-ridden legislature was dead set against passenger subways, Beach intended to surprise the public with a model railway which would generate such overwhelming popular support for a full-scale subway that no politician would dare block it.[31]

The subway would run up Broadway for one block, from Murray Street to Warren Street. This route was selected because it was close to the site of the new post office being planned at the corner of Broadway and Park Row. By carefully maintaining his parcel dispatch cover story, Beach kept his real objective a secret. Like Edgar Allan Poe's purloined letter, the subway was hidden in plain view; in fact, the route was actually within sight of city hall. New York's pols slumbered through January and February 1870, blissfully unaware of what was happening below their feet.

Beach devised a hydraulic tunneling shield to expedite the project. An improvement on a shield that Sir Marc Brunel had designed forty years earlier to burrow under the Thames, Beach's invention was important for enhancing worker safety but was also significant for accelerating the digging. With the use of Beach's shield the tube could be moved ahead eight feet on a good night.

Working conditions were nevertheless miserable. The tunnel measured only eight feet wide and did not contain nearly enough room for the people who had to toil there. For the eight-man crews, the tube was a living hell of hot, stinking air and dim, flickering lantern light. The sound of the horse cars and omnibuses clattering along Broadway, twelve feet above the roof of the tunnel, heightened their claustrophobia. Despite this adversity, the project proceeded on time. The only serious delay occurred when the miners encountered the foundation wall of an old Dutch fort. One after another the stones had to be removed and passed back through the tunnel.[32]

The heart of the Beach Pneumatic Railway was a huge rotary blower that was dubbed the Western Tornado. Manufactured by the P. H. & F. M. Roots Company of Connersville, Indiana, this fifty-ton ventilator was so massive that a special train had to be assembled to carry it to New York. It worked simply enough: Installed in a hidden recess at the Warren Street end, the Western Tornado produced one hundred thousand cubic feet of air per minute that would drive the car down to Murray Street at about ten miles per hour. When the car reached Murray Street, its wheels tripped a telegraph wire that sent a signal to an engineer at Warren Street, who adjusted the valves to reverse the air current. The ventilator then sucked the car back to its starting point. The company fitted the stations with vacuum-tight doors to ensure that these sudden changes in air pressure did not disturb passengers.[33]

Beach splurged on lavish decorations for his subway. In order to dramatize how ugly the horse railways and omnibuses were, the Beach Pneumatic Railway was designed to look more like an upper-middle-class home than an ordinary public conveyance. Oil paintings, settees, chandeliers, a grandfather clock, mirrors, a fountain, and a goldfish tank adorned the waiting room. The little car, which was half the size of a horse car and accommodated twenty-two people, featured richly upholstered seats and zircon lights. Even the plain interior walls of the tunnel were neatly whitewashed in order to present a clean, sanitary appearance.

Unveiled on February 26, 1870, the Beach Pneumatic Railway instantly became a popular sensation. It was a dazzling spectacle, conveying a futuristic vision of transport progress. According to the *New York Herald,* the opening revealed an "Aladdin's cave" full of hidden magic.[34] Tens of thousands flocked to the new wonder, gladly paying twenty-five cents for the novelty of an atmospheric car ride; over four hundred thousand people rode it that first year. To New

Yorkers who were weary of bone-breaking horse omnibuses and railways, the gentle motion of compressed air was a revelation. "It was propelled like a sailboat before the wind," marveled *Appleton's Journal*.[35]

Despite the unqualified public success of his demonstration model, Beach failed to achieve his primary objective of building a full-scale subway. The major obstacle was not political opposition, however. A startling *New York Times* exposé of municipal bid-rigging, kickbacks, and faked leases that was published in July and August 1871 led to the ouster of the Tweed ring the following year, paving the way for the legislature to pass a charter for Beach's railway on April 9, 1873. But even after Albany gave the go-ahead, Beach still had to overcome significant technical and financial barriers.[36]

Pneumatic propulsion was one problem. Compressed air proved to be satisfactory for the limited purpose of transmitting letters and parcels, but it was utterly impractical for a comprehensive passenger subway.[37] The cost of producing the tremendous volume of air needed to move the trains was prohibitive because a great deal of energy was wasted in transmission. In addition, pneumatic power could not be controlled easily and lacked the flexibility needed to adjust speeds, load and unload cars, and run multiple trains. Atmospheric power was a fascinating toy, but it was not the answer to the search for better propulsion.

By the time he received his charter in 1873, Beach had abandoned compressed air for steam power. This was a wise move. Even though steam locomotives spewed sparks, cinders, and ashes that would have annoyed passengers and crews, engineers could ease this environmental hazard. For instance, the London Metropolitan was constructed close to the surface of the street, permitting the enclosed portions of the tunnel to be interspersed with open cuts that increased air circulation. Steam certainly had flaws, but it offered a dependable, safe, and proven source of energy and could have been used in a New York subway. In switching to steam Beach cleared away the last significant technological roadblock standing in the way of his dream.[38]

But Beach was unable to raise money for construction. Economics, not technology, was the primary obstacle. Without any hope of government subsidies to offset the extraordinary high cost of subterranean building, Beach could not persuade financiers to shoulder the entire burden of this risky enterprise. His hopes were completely dashed when the disastrous stock market panic of September 18, 1873, precipitated a severe financial depression that contracted the supply of private investment dollars.[39]

In 1874, Beach admitted defeat and suspended the pneumatic railway operations. He rented the tunnel to a series of private businesses in an effort to recoup his financial losses; over the years it served as a shooting gallery, a wine cellar, and a storage room for a clothing store. The tunnel was finally abandoned, sealed, and completely forgotten until February 1912 when construction workers from the Degnon Construction Company rediscovered it. While digging a new subway for the Brooklyn Rapid Transit Company, they broke through the brick wall and entered the old chamber. Using candles to light the way, these laborers saw that Beach's showcase remained in surprisingly good shape; although the wooden car had almost completely rotted away, the hydraulic shield and the waiting room were well preserved. Unfortunately, this archaeological treasure was located on the site projected for the Brooklyn Rapid Transit Company's city hall station, and it would have to be demolished. Before the wreckers began destroying the landmark, however, the New York Public Service Commission rescued the tunneling shield. Alfred E. Beach's son, Frederick, an editor at *Scientific American,* gave the shield to Cornell University, which put it on display in Sibley College. That relic became the only surviving artifact from New York's earliest subway.[40]

The Elevated Railways

New York's first rapid transit lines were elevated rather than underground railways. Because they were much cheaper to build than subways, els, as they were known, did not require government subsidies and could be constructed within the laissez-faire economic framework of the day. At the same time Alfred E. Beach was struggling to finance his pneumatic tube, the first elevated lines were already inching up the island.

Charles T. Harvey, a Connecticut-born inventor, opened the first elevated line in the world on July 3, 1868, on a half-mile stretch of Greenwich Street from Dey Street to Cortlandt Street in lower Manhattan. Although Harvey dreamed of building a vast twenty-five-mile route stretching across the Harlem River to Kingsbridge and Yonkers (hence the company's optimistic name, the West Side & Yonkers Patent Railway), his original line was a crude, jerry-built contraption that did not inspire much confidence. It consisted of one track supported by a single row of slender iron stanchions running along the curb. During its early trials some observers feared that at any minute the car would topple off the track and crash to the street. "It was a rather shabby and frail-looking structure," *Appleton's Journal* noted ten years later, "and was mockingly called, by the people who refused to ride upon it, the railway upon stilts, or the one-legged railway."[41]

The West Side & Yonkers employed a system of cable propulsion that also failed to inspire confidence. Power was generated by stationary steam engines located in underground vaults; they pulled a three-quarter-inch steel cable through a wooden box that was situated between the rails. To go forward, the driver engaged a clamp that extended through the bottom of the car and gripped the continuously moving cable; to stop, he released the lever. The cable was delicate and frequently broke, but the West Side & Yonkers attained a top speed of fifteen miles per hour and was deemed a success. Despite its flaws, Charles T. Harvey began to extend his railway up Greenwich Street and Ninth Avenue. A year later the railway was completed as far as the Hudson River Railroad's terminal at Thirtieth Street.[42]

Then on September 24, 1869—Black Friday—the stock market collapsed. Harvey, caught short of capital, was ruined. The West Side & Yonkers suspended operations and declared bankruptcy; on November 15, 1870, the company's property was sold at a sheriff's auction for the pathetic sum of $960. The following October the line was reorganized as the New York Elevated Railway. The new owners proposed a 160-mile network of new lines extending to northern

Westchester County; they replaced cable propulsion with more reliable steam locomotives and resumed construction of the Ninth Avenue route.[43]

Meanwhile, a rival elevated line appeared. New York's second overhead was the brainchild of Dr. Rufus Gilbert, a New York City surgeon who was renowned for having performed several difficult operations and who had supervised the U.S. Army hospitals during the Civil War. On a tour of European hospitals in the late 1850s Gilbert became fascinated with the problem of providing rapid transit for large cities. After the war he abandoned the practice of medicine and joined the search for a motive power for urban mass transit. To Gilbert this was not a career change; he saw rapid transit as an instrument for promoting public health. In his view rapid transit would empty American hospital beds and reduce the high urban death rates by allowing the poor to move from the overcrowded, disease-ridden slums to the green hills of the surrounding countryside.

Rufus Gilbert rallied to Alfred E. Beach's favorite cause by championing a pneumatic elevated railway. But although he received a franchise for a west side route in June 1872, Gilbert could not raise the money for construction. His fate bore an eerie resemblance to Charles T. Harvey's. Like Harvey, Gilbert lost control of his company after a stock market crash (in his case, that of September 1873). Like the New York Elevated, the new management of the Gilbert Elevated switched to steam power and announced plans for new routes.[44]

As the depression of the 1870s eased and investor confidence returned, both New York Elevated and Gilbert Elevated were eager to implement their building programs. To do so required legislative approval, however. In response to the Gilbert company's plea, the state legislature passed the Husted Act of 1875, a landmark measure that set the standard for rapid transit development for two decades. Under the Husted Act the mayor of New York City had the authority to appoint a five-man rapid transit commission, which was empowered to lay out routes and assign a franchise to a private operator who would build, equip, and operate the facility.

The Husted Act limited government's role in the economical aspects of rapid transit to the bare minimum. Municipal participation involved little more than deciding whether or not to approve a proposal submitted by a company. The government was passive, acting only in response to private initiatives, rarely putting forward its own ideas, and never troubling to monitor the railway's actual performance once the franchise was awarded. Indeed, these rapid transit commissions were conceived as temporary panels of distinguished citizens rather than as permanent government bureaus with large, expert staffs and extensive regulatory powers; according to the act, the commissions were to go out of existence one month after selecting the private corporation that would operate its routes, without even waiting for construction to be completed.[45]

Formed on July 1, 1875, the Rapid Transit Commission of 1875 made its awards two months later. The Gilbert Company (soon renamed the Metropolitan Elevated Railway) acquired routes on Second and Third avenues, while New York Elevated (which was already operating Harvey's original Ninth Avenue line) obtained a route on Sixth Avenue. On September 1, 1879, the Metropolitan and New York Elevated were consolidated into a single gigantic corporation, the Manhattan Elevated Railway Company.[46]

Construction of the Manhattan company's four routes moved very quickly. By August 1880 the Ninth, Second, and Third Avenue els reached the Harlem River. A vast new area of northern Manhattan was now open.[47]

To the Top of the Island

Not long afterward, on a bright day in 1881, Iza Duffus Hardy, an Englishwoman, and two young friends decided to take a sightseeing trip on the "open-air el" to northern Manhattan.

They climbed the steep steps to the Fiftieth Street station of the Sixth Avenue elevated, paid ten cents each for their tickets, and boarded a smart-looking uptown train.[48] The Sixth Avenue, which ran from South Ferry to Central Park, was New York's best overhead route. Their car was light, airy, and extremely comfortable; its wooden seats were cushioned, and its floors were covered with braided mats.

Up Sixth Avenue, around the sharp curve at Fifty-fourth Street where the Sixth Avenue route connected with the Ninth Avenue line, and then up Ninth Avenue, the steam locomotive pulled the train past the region of shops and homes and into the unsettled territory of the west Sixties and Seventies. The West End, as this area was known at the time, was expected to become a choice residential quarter, but in 1881 the streets were little more than numbers, and most of the land was vacant. The entire district had a shabby, forlorn appearance. In 1884, Singer Sewing Machine heir Edward S. Clark erected a luxury apartment building at Seventy-second Street and Central Park West, and it was dubbed the Dakota because it was so far out of town that it might as well have been in the badlands.[49]

Whirling up Ninth Avenue, the train passed a shantytown at 100th Street. Mockingly called "Irish villages" or "goat towns," these poor communities were scattered across the metropolis, squeezed onto fringe land that nobody else wanted yet. Hardy would later write:

> A whole tribe of immigrants have "squatted" on a broken ground, all holes and hillocks and blasted rocks. A horde of dwellings of all sorts and conditions, from the mere mud hovel to the neat little wooden cottage, have sprung up, straggling anyhow and everywhere, crowding together or scattered apart, one thatch-roofed hut perched up high on the edge of a huge rock, looking as if a push would send it over—one dwarfish cabin nestling on the leeside of a big block; another has ingeniously utilised an angle of rock for two of its walls; some have laid out little slips of garden, still gay with autumn flowers. Garments of all kinds are hung flapping in the breeze to dry; a brood of ducks are dabbling in a muddy pool; geese and fowls, gaunt pigs and bare-footed children, all run wild together.[50]

This squatter's camp would vanish within fifteen years, a casualty of the new apartments and homes constructed there.

At 110th Street the locomotive entered one of the wonders of New York: a grand serpentine curve where the route descended from the island's western ridge line to the Harlem lowlands. "Round the first sharp curve our train curls like a snake," Hardy said. "Looking out from the last car, we can see the engine seemingly running back upon us, and the whole train apparently doubled in half." At its highest point the iron trestle rose a spectacular sixty feet above the ground, providing Hardy and her traveling companions with a great view of Central Park and Harlem village. New cross streets, raised above the natural level of the land, had been opened across the fields and meadows; a few homes and inns had already cropped up there.

Hardy's train stopped at the 155th Street terminal. The area surrounding this depot was a lively recreational center where picnic groves, beer gardens, hotels, and dance halls enticed day-trippers who wanted to escape from the city. Bypassing these honky-tonk establishments, Hardy and her friends took a train one stop across the Harlem River, to Highbridge, in what is now the Bronx. After leaving the station the three strolled down a country lane into the depths of the woods. This landscape of hickory and oak presented a stark contrast to the redbrick and brownstone city at the other end of the el. "We are utterly out of sight and hearing of the world as if the big, busy, booming city lay a day's journey off."[51]

The elevated railway, which went about twelve miles per hour, dramatically reduced traveling time. Iza Duffus Hardy had gone from West Fiftieth Street to Highbridge in only about forty-five minutes; if she had started from somewhere in lower Manhattan rather than West Fiftieth Street, her journey to Highbridge would have taken an hour and a half. By horse-powered coach these trips would have taken at least three times as long.

But the elevated railway was a mixed blessing for the city. Although the els dramatically increased travel speeds, many New Yorkers complained that their construction lowered the quality of neighborhood life. Their bulky girders, columns, and rails blocked the sunlight and kept the surface of the streets in perpetual twilight. Moreover, the els were very dirty. Visiting Manhattan in 1889, Australian James F. Hogan described the coal-burning elevateds as an "ever-active volcano" that belched thick black smoke, ashes, cinders, and sparks; in his view these "unsightly and abominable" eyesores constituted an "aerial nuisance" that caused "the permanent disfigurement of a great metropolis" and posed "a severe trial to the average nervous system."[52] English traveler William G. Marshall had good reason to agree with Hogan's evaluation. One day in 1880, while walking across the intersection of Sixth Avenue and Twelfth

Street in Greenwich Village, Marshall was splattered by axle-box oil that fell from a passing el train; although he rushed to a nearby barbershop and tried to clean up the mess, his hat, coat, and necktie were ruined. (Fortunately, Marshall had a knack for seeing the silver lining in every cloud: Since he had been gazing upward at the time, he considered himself "lucky . . . my mouth [was not] open" when the oil hit.)[53]

In addition to being filthy, the els were extremely loud. According to novelist William Dean Howells, New Yorkers who lived or worked near the railways had to endure up to "nineteen hours and more of incessant rumbling day and night." The roar of the trains assaulted residents in the shops and on the streets and invaded their flats, disturbing their sleep. The sound, Howells added, was particularly bad on streets where elevated trains, horse cars, wagons, and carriages competed for space:

> No experience of noise can enable you to conceive of the furious din that bursts upon the sense, when at some corner two cars encounter on the parallel tracks below, while two trains roar and shriek and hiss on the rails overhead, and a turmoil of rattling express wagons, heavy drays and trucks, and carts, hacks, carriages, and huge vans rolls itself between and beneath the prime agents of the uproar. The noise is not only deafening, it is bewildering; you cannot know which side the danger threatens most, and you literally take your life in your hand when you cross in the midst of it.[54]

Notwithstanding these environmental hazards, however, the primary effect of the elevateds was to open new lands for settlement and to stimulate a construction boom above Forty-second Street. Tenements and brownstones were erected in the corridors along the routes of the Ninth, Sixth, Third, and Second Avenue railways during the 1880s. The island's growth was incomplete and uneven, however. Although buildings covered much of the available ground on the Upper East Side by 1890, thousands of lots remained vacant on the Upper West Side and in Harlem. Northern Manhattan was no longer the bucolic landscape that had charmed Iza Duffus Hardy in 1881, yet it was not completely urbanized, either.[55]

In 1890, New York City had the largest and most comprehensive mass transit system in the world. As mentioned earlier, New York had more total mileage than London, which contained nearly three times as many residents. Transit was thus extremely important in New York City. The 1890 census reported that New Yorkers averaged almost three hundred mass transit trips every year, compared to only about seventy-four rides for each Londoner.[56]

Even this huge transit network was inadequate. The elevated railways lacked the carrying capacity and high speeds demanded by the city's rapid population growth and its difficult geography. The els were already overcrowded by 1890, and the construction of new routes or extra storage yards was precluded by space limitations, especially in the financial district on the island's southern tip. Similarly, the elevated railways were not fast enough to kindle residential growth in the new areas at the northern end of this long, narrow island. Above Fifty-ninth Street far more land remained vacant than was developed.

The solution, as Abram S. Hewitt urged the Board of Aldermen in his landmark message of January 31, 1888, was an underground railway.

2

MAKING GOVERNMENT SAFE FOR BUSINESS

A Traditional Approach

On January 31, 1891, Governor David B. Hill signed into law the Rapid Transit Act of 1891. This statute, signed three years to the day after Abram S. Hewitt's address to the Board of Aldermen, began a drive toward realizing Hewitt's dream of a great urban railway.[1]

Although the new law affirmed Hewitt's argument about the importance of rapid transit to New York's future as a world city, it shunned his key recommendation for public investment in a subway. Only a few of the city's politicians and businessmen gave serious thought to implementing Hewitt's policy proposal. Most opposed municipal financing as an encroachment on the rights of property and as an open invitation to Tammany thievery. Two decades earlier the regime of William M. Tweed, the boss of the Tammany Hall Democratic organization, plundered more than $100 million from the city through kickbacks, fraudulent leases, and fake bills. In the 1890s memories of the Tweed ring remained so strong that private, not municipal, ownership was widely regarded as the best safeguard of the public treasury. Fear of corruption thus reinforced the general laissez-faire thinking of the day. Consequently, the Rapid Transit Act of 1891 authorized the formation of a rapid transit commission that would lay out routes for an underground or elevated railway and then assign a franchise for construction and operation to a private company. In the tradition of the Husted Act of 1875, the new commission acquired only enough power to promote corporate development of rapid transit and could neither build nor run the line itself.[2]

One critic of this approach was Jacob H. Schiff, the suave, tough-minded senior partner of Kuhn, Loeb & Company and a distinguished immigrant German Jewish leader. Schiff was convinced that private subway financing simply would not materialize. At a hearing in March 1891 he warned New Yorkers not to assume that investment banks would fund the construction of an underground railway just because they had underwritten the elevateds two decades earlier. According to Schiff the higher cost of a subway—perhaps as much as $50 million or even $100 million—meant that no banker in his right mind would dare tackle the project. "There then appears to exist only one way in which rapid transit can be secured promptly and economically," Schiff asserted, "and that is by putting the plans . . . into execution at the expense of the city."[3]

But nobody paid any attention to Schiff's warning, and rapid transit planning proceeded along its old path.

The Steinway Commission

The new rapid transit commission was chaired by William Steinway, a native of Seesen, Germany, who had emigrated to the United States as a fourteen-year-old in 1850. Three years later Steinway helped his father and two older brothers start a piano manufacturing business in a rear building at 55 Varick Street in downtown Manhattan.

William Steinway deserves much of the credit for transforming this fledgling company into Steinway & Sons, perhaps the foremost piano manufacturer in the world. A natural leader, a man of immense personal charm, meticulous, hardworking, and somewhat intimidating, William Steinway adjusted easily to his adopted country. Unlike the older males in his family, he learned to speak English well and always felt at home in America, cultivating a wide circle of "American'' bankers, merchants, and lawyers who served as useful contacts. He also mastered American business practices, including the art of marketing. With astute advertising, Steinway helped give the piano an aura of prestige and glamor that appealed to the respectable middle class, vastly increasing its sales and turning the instrument into a cultural icon.

William Steinway entered the real estate business as a result of his efforts to protect Steinway & Sons from the labor militancy that gripped Manhattan. After a series of strikes rocked its factory at Fourth (now Park) Avenue and Fifty-second Street in the late 1860s, the company bought a stretch of land that fronted on Bowery Bay in Astoria, Queens, then a sparsely settled expanse of salt marshes, meadows, and woodlands. There, across the river from turbulent Manhattan, Steinway & Sons built a sprawling factory and company housing for its workers.

Afterward William Steinway started to boost Astoria as a residential dormitory of Manhattan. His activities fit a classic description of the real estate promoter: He accumulated more property, graded and leveled the land, subdivided it into lots, laid out streets, dickered with local politicians for municipal services, and advertised his parcels for sale. To make this remote section of Queens more accessible to homeowners, Steinway operated a horse railway line that shuttled to the East River ferry, the Steinway & Hunter's Point Railroad Company. By 1890 more than fifteen thousand people lived in Astoria.

William Steinway's career as a developer was the major reason Mayor Hugh J. Grant named him to the transit commission. At a time when the modern notion of ''conflict of interest'' was completely unknown, Steinway was attractive because of his business connections, not in spite of them. To his contemporaries Steinway's first-hand transit experience was a valuable source of expertise about rapid transit that the tiny government bureaucracy otherwise would not have, while his personal rectitude promised a high level of honest and faithful service. This sort of curriculum vitae was a common qualification for government. Like Steinway, the other four commissioners were prominent businessmen with strong commitments to urban growth: John Starin, owner of a fleet of tugboats and lighters that plied New York Harbor; Samuel Spencer, president of a southern railroad; Frederick Olcott, president of the Central Trust Company; and railroad attorney Eugene Bushe.[4]

The Steinway Commission, as it was called, issued a plan for a subway in October 1891. There would be two routes: a main line running from South Ferry up Broadway to the Bronx, and a branch railway forking off from the main line at Union Square and going up Madison Avenue and across the Harlem River, also to the Bronx. After Mayor Grant, the Board of Aldermen, and other local authorities approved this proposal, the Steinway Commission drafted a contract for the railway's construction and operation.[5]

The sale of this franchise posed a critical test of the commission's laissez-faire approach. In a series of upbeat newspaper interviews, William Steinway insisted that the contract offering would attract a throng of investors. He bragged to one reporter that the underground railway was already as good as built:

> When this [plan] is accomplished, merchants and their clerks who now stand and sit about downtown restaurants for their lunches will jump aboard these cars and get home in five, ten, and fifteen minutes for luncheon with their families. . . . People who now live over in Jersey or out in Queens County on Long Island will come to reside in the upper part of New York, as nearer their business and as a more desirable place to live.[6]

These were empty words, however. Privately, Steinway suspected that nobody would bid for the line. He was right. When, on December 29, 1892, the commission tried to auction a 999-year contract for their subway, not a single serious bid was tendered. The only proposal of any kind was an absurd offer (which the commission immediately rejected) from a businessmen named W. Nowland Amory who wanted to pay just $1,000 in cash for these valuable privileges. Jacob H. Schiff's prediction had come true: No financier would touch the subway, especially in a weak economy.[7]

This failure put the Steinway Commission on the defensive. The *New York World,* the *New York Times,* and other papers grumbled that Steinway, Starin, Bushe, and the others were no closer to achieving a subway than the day they had become commissioners.

The commission also felt pressure from a group of real estate speculators who had bought large tracts of land along the projected route in Harlem and Washington Heights. Instead of profiting from rising property values, these developers were now stuck with millions of dollars' worth of acreage that they could not unload. To avoid financial ruin, these realtors urged the Steinway Commission to build a rapid transit railway to upper Manhattan right away. The Real Estate Exchange and the *Real Estate Record and Builders Guide,* an influential trade journal, seconded their demands.[8]

Consequently, the Steinway Commission decided to abandon its subway project and try for an elevated railway. Because an elevated railway could be completed more quickly and more cheaply than an underground, it offered a way of silencing the commission's critics and providing rapid transit within the framework of existing law and laissez-faire doctrine. But even though an el would have resolved the commission's immediate political problems, its construction would have killed Abram S. Hewitt's vision of first-rate urban transport. A slow-moving, low-capacity el could not overcome New York's stubborn geographical barriers or quicken the growth of the northern districts or accommodate future traffic increases. Chairman William Steinway, who initially held out for a subway and tried to dissuade the others from voting for an el, sensed the danger: "To my dismay," he confided to his diary on January 14, 1893, "I see that I stand alone in my stand to guard the city from being further disfigured on [the] streets. I feel dreadful."[9]

There was a more serious danger. This decision played into the hands of the Manhattan Railway Company, a powerful corporation assembled a decade earlier by Jay Gould, the notorious financial manipulator who first attracted national attention when his scheme to corner the gold market triggered the "Black Friday" stock market panic of September 24, 1869, and who then gained control of the Associated Press news service, the Western Union telegraph, and the *New York World* newspaper. In 1881, Gould set his sights on Manhattan's two elevated railways, the Metropolitan and the New York companies. Using legal muscle, political bribery, and financial chicanery, Gould snatched the two lines and merged them into a single gigantic enterprise called the Manhattan Railway Company that monopolized the island's entire rapid transit network. Under Gould's management the Manhattan became infamous for its slovenly passenger service, stock manipulation, and closeness to Tammany Hall. After Gould died in December 1892, his son, George J. Gould, and robber baron Russell Sage took over and perpetuated his public-be-damned attitude.[10]

The Steinway Commission was so hungry for success that it decided to ignore the Manhattan's unsavory reputation. By April 1893 the Steinway Commission began courting George J. Gould and Russell Sage with a proposal to build a new elevated railway. These talks dragged on for months, perhaps because Gould and Sage were not serious about reaching an agreement for a new elevated. They were probably trying to coerce the commissioners into authorizing improvements for their existing routes. Gould and Sage had a monopoly over the island's rapid transit lines. Had the talks been successful, they would have retained their monopoly and acquired enough power to veto future transit projects. But in June 1893 the stock market crashed, ushering in the severe financial depression of the 1890s.[11]

The stock market crash exposed the weaknesses of the existing political system and created a strong impetus for change; it was clear by early 1894 that private financing of a subway was very unlikely. Consequently, the leadership of the Chamber of Commerce of the State of New York decided to initiate government investment in rapid transit—a very special type of business-dominated government investment, that is.[12]

The Chamber of Commerce

A banker named Richard T. Wilson was the first to advocate that the Chamber of Commerce should lobby for public investment in rapid transit. Wilson was an old hand at uniting business and government behind a great civic cause, although his highly unorthodox experience was not what Abram S. Hewitt or the other members of the commercial aristocracy had in mind for their subway. A native of Habersham County, Georgia, who was working as a merchant in eastern Tennessee at the start of the Civil War, Wilson promptly enlisted in the military effort. Applying his knowledge of business on behalf of the rebellion, Wilson served first as commissary general of the Confederate army and then as fiscal agent for President Jefferson Davis's government in London. After the war Wilson moved to New York and founded an investment house, R. T. Wilson & Company, that emerged as one of Wall Street's foremost banks. As Wilson became a very rich man, the inconvenient little matter of his wartime treason was discreetly forgotten, and his family was absorbed into New York's aristocracy through membership in the Union, Knickerbocker, and Turf and Field clubs, racing at Saratoga and sailing at Newport, and marriage into the Vanderbilt, Astor, and Goelet clans.[13]

Richard T. Wilson had an excellent reason for championing municipal financing of a subway: He expected the money to go to him. Wilson had approached the Steinway Commission in 1893 with a proposal for an underground railway to be partially funded by a government loan. According to Wilson's scheme the municipality would cover two-thirds of the total construction cost of $45 million through a loan to R. T. Wilson & Company, while the bank would raise the remaining $15 million from private sources. The Steinway Commission had originally rejected this proposition, but the tables turned by January 1894 and the commission apparently invited Wilson to resubmit his proposal. He did.

Wilson's idea was extremely controversial. Although the Steinway Commission okayed the plan and forwarded it to city hall for review by municipal officials, opponents branded this scheme to entrust a private company with tens of millions of city dollars as an irresponsible folly that bordered on criminality. The most serious blow came on January 31 when Mayor Thomas F. Gilroy rejected it.[14]

The next day Richard T. Wilson took the floor at a biweekly meeting of the Chamber of Commerce and delivered an impassioned plea for help.[15] Wilson had come to the right place, for his business audience strongly supported a subway. To the Chamber of Commerce, rapid transit was the solution to two problems that vexed New York City at this time. As Abram S. Hewitt had pointed out eight years earlier, the first problem involved the urgent need for the city to gain an edge over rival seaports through residential settlement of the Upper West Side, Harlem, and the annexed district, and through the broadening of the city's tax base. The second problem concerned Manhattan's horrendous street traffic. The tall buildings being erected to house Wall Street's new banks, insurance companies, and corporate headquarters were pouring more and more pedestrians and lorries onto streets already swollen to the bursting point. As a result wholesalers, department store merchants, and garment manufacturers complained that their shipments were delayed and that their employees did not arrive at work on time. The answer was a subway.[16]

The Chamber therefore appointed a special committee to explore Wilson's idea, chaired by one of the Chamber's most distinguished members, an Irish Protestant immigrant produce merchant named Alexander E. Orr. By virtue of this committee assignment, Orr would figure prominently in rapid transit planning for over a decade.[17]

Two weeks later, on February 15, Orr's committee returned with a highly favorable report that compared the projected subway to another magnificent public work that had also benefited from a government subsidy: the transcontinental railroad. Hundreds of millions of federal land-grant dollars had carried the tracks of the Union Pacific and the Central Pacific railroads across the Great Plains and through the Rocky Mountains from Nebraska to California. Opened on May 10, 1869, with the famous golden spike ceremony at Promontory Point, Utah, this railroad redeemed its generous subsidy by peopling the trans-Mississippi west and by tying the Pacific Coast to the eastern seaboard. The parallel between the transcontinental railroad and the underground railway was striking. Claiming that "the future of this city as the commercial metropolis of the United States . . . requires the very best system of rapid transit," the Orr committee advised the Chamber to sponsor Richard T. Wilson's program.[18]

But Abram S. Hewitt and Jacob H. Schiff denounced the committee's report before the members had a chance to vote on it. Hewitt, who took the lead in the debate that followed, declared that the committee's own transcontinental railroad example actually showed the perils of bankrolling a private corporation, rather than the advantages. The ex-mayor caustically reminded his listeners—and the Orr committee—that the railroad had precipitated the worst scandal in the history of the federal government, the notorious *Crédit Mobilier* affair. Union Pacific stockholders had stolen $30 million of U.S. Treasury bonds in that celebrated incident.

Hewitt's harsh comments struck a nerve. With Tammany Hall spoilsmen apparently lurking in every dark corner of the city government, these executives were concerned that the subway project would turn into a bonanza of sweetheart deals and kickbacks. According to Hewitt and Schiff, the best way to deter corruption was by replacing Wilson's plan with Hewitt's formula for public ownership and private management. They argued that municipal ownership was a sounder and more conservative policy than a loan because it would ensure stricter public supervision of the companies.[19]

As old lions whose advice commanded enormous respect, Abram S. Hewitt and Jacob H. Schiff easily persuaded the Chamber to drop Wilson's scheme, which it did on March 1, 1894. That same day the Chamber of Commerce instructed the Orr committee to prepare a legislative bill incorporating the Hewitt formula. Within the next two weeks Orr arranged for such a measure to be drafted, dispatched to Albany, and introduced into the state legislature.

Not surprisingly, this Chamber bill was strongly anti-Tammany. It was supported by the Republican leadership of the state senate and the assembly, who packaged it with a number of other measures intended to declaw the Tammany tiger. During the legislative debates over the rapid transit initiative, GOP representatives made hoary references to Boss William Tweed (who had been rotting in his grave for over fifteen years).[20]

In this fight against the Tammany machine, however, the Chamber of Commerce revealed its disdain for electoral politics. Attorney Henry R. Beekman, the principal drafter of the bill, did not think that either party should have control over New York City rapid transit.[21] Like Hewitt, Schiff, and Orr, Beekman thought responsibility for a subway should be entrusted to a special breed of honest, upright, farsighted men—namely, themselves. They sought to remove rapid transit decisions from normal administrative channels and to invest the mercantile elite with complete authority for the subway. In effect they wanted the Chamber of Commerce to control the new transit system.

They would accomplish this by holding five of the eight seats on the new rapid transit commission. Five businessmen commissioners were identified by name in the act; the other commissioners, all *ex officio,* were the president of the Chamber of Commerce, the mayor, and the city controller. This elite domination would be perpetuated by a provision that allowed the majority of the board to fill its own vacancies without the participation of any outsiders, not even the governor or the mayor.[22]

The Chamber of Commerce's initiative won the support of mercantile organizations such as the Board of Trade and Transportation, the Cotton Exchange, and the Fruit Exchange; the Real Estate Exchange and the *Real Estate Record and Builders Guide;* the City Club, the Good Government Club, and other groups. Introduced during a period of commanding Republican legislative majorities, the bill swept through the assembly by a margin of ninety-three to three and the senate by twenty-eight to zero. Governor Roswell P. Flower signed the Rapid Transit Act of 1894 into law on May 22.

The bill had been opposed by many of New York City's trade unions and by a handful of genteel political reformers who, though too weak to defeat it, nonetheless succeeded in pushing through an amendment designed to curb mercantile power. This challenge was masterminded by Charles B. Stover, a Pennsylvania-born graduate of Union Theological Seminary who was a dynamic settlement house worker and political activist on the Lower East Side. Acting on behalf of more than one hundred unions, Stover argued that the people should not be completely excluded from a decision that would reshape their city. Stover proposed that a referendum be held at the next general election so that the electorate could have an opportunity to either accept or reject the Hewitt formula of municipal ownership and private construction and operation.

The Chamber's leadership regarded this union recommendation as a serious threat to business control. But even though Abram S. Hewitt and Henry R. Beekman went to Albany to lobby against it, not even the most stalwart Republican legislator wanted to go on record as a foe of democratic participation, so Stover's amendment passed. But in November 1894, New York City voters trooped to the polls and sanctioned the Hewitt formula by a resounding vote of 132,647 to 42,916. By popular acclaim, business reigned supreme.[23]

To the IRT

The Board of Rapid Transit Railroad Commissioners (RTC) was eager to succeed where the Steinway Commission had failed. But it encountered stiff opposition immediately.

The first resistance came from an unexpected quarter: the real estate industry. On February 16, 1895, the RTC adopted a subway plan that laid out one route from South Ferry up Broadway to the Bronx and another route from Union Square under Park Avenue to the Bronx. This plan was almost identical to the Steinway Commission's 1891 map.[24] There was, however, one small but critical change. In order to satisfy a provision of the 1894 act that stipulated construction of the subway must cost less than $50 million, the commission changed a section of the route that ran under Broadway between Ninety-second Street and 124th Street from an underground to an elevated railway. To the RTC this switch to a cheaper but uglier el offered a means of saving $2 million without causing any environmental damage to an established neighborhood; this twenty-two-block stretch of the Upper West Side and Morningside Heights was full of empty lots.

But the RTC had not calculated on real estate speculators such as Francis M. Jencks, Charles T. Barney, Clarence True, and Cyrus Clark, who had acquired much of this property in anticipation of the subway's opening. Afraid that the blighting of Broadway would devalue these holdings, Francis M. Jencks mounted a campaign to reverse the decision. The highpoint of this angry protest occurred on March 12 when more than fifty developers invaded a meeting of the RTC on the ninth floor of the Home Life Building at 256 Broadway and created an uproar. One agitated realtor accused the commissioners of being mentally ill and threatened to beat up every single one of them. Another owner tried to play on the RTC's sympathies by acting like a plain householder whose life savings were about to be snatched away by a merciless and unfeeling government; this man—in reality a rich business executive and prominent Chamber of Commerce member—complained that he lay awake night after night worrying about the fate of his "little property."[25]

The protest worked. The commissioners not only esteemed the speculators as political allies and as business acquaintances but also shared their goal of encouraging the development of upper Manhattan. Apologizing profusely, the commissioners hastily canceled the el and revised their map.[26]

The RTC was constrained by a review process designed to avert wasteful spending on public works. The RTC's plan had to be approved by the mayor, the Board of Aldermen, and either local property owners or the state courts before the commission could award a franchise to a private contractor.[27]

The first stage was easy. Mayor William L. Strong and the Board of Aldermen approved the proposal in May 1895. The next step compelled the RTC to obtain consents from the owners of one-half of the total assessed value of property that abutted the projected route. In effect the RTC had to conduct a special referendum where only the people and the corporations who owned property along the subway route could vote and where each voter's "ballot" was weighted in direct proportion to the monetary value of his holdings. It was a vivid expression of the rights of property.

The RTC failed to secure the needed consents. The greatest opposition came from titleholders in downtown Manhattan who were concerned that the digging of a subway would undermine the foundations of their big buildings and obstruct street and pedestrian traffic. Unlike the Upper West Side speculators who eagerly awaited the underground (as long as it stayed underground), these downtowners worried that a subway would damage their investments.

But the law provided the RTC with a recourse. After the property owners rejected its plan, the commission petitioned the general term of the New York State Supreme Court in the fall of 1895 to appoint a panel of three citizens who would determine whether the plan was fiscally sound or not. If the panel certified the plan and if the appellate division of the supreme court concurred with the panel, then the statute allowed the RTC to go ahead and assign a contract. After lengthy legal delays, the three-man panel solidly endorsed the subway on March 6, 1896.[28]

When the Rapid Transit Commission asked the appellate division to confirm this report, Presiding Justice Charles H. Van Brunt and Justice William Rumsey delivered a scathing decision on May 22, 1896, which focused on the statutory requirement that the subway cost less than $50 million to build. Contemptuously dismissing the RTC's $30 million estimate as a fantasy that betrayed "utter ignorance of the [true] cost," Van Brunt and Rumsey characterized the subway as a bloated white elephant that not only would exceed the legal limit but would also inflate the municipal debt load and mar the city's credit ranking.[29]

This ruling plunged the commissioners into despair. "It means the end . . . to the plans we have presented," RTC president Alexander E. Orr said. "There can be no appeal."[30] In a last-ditch strategy to keep its quest for rapid transit alive, the RTC entered into negotiations with the Manhattan Railway Company for the construction of an elevated railway. Suddenly, George J. Gould and Russell Sage were in the picture again.

This development horrified William Steinway, who blamed the Manhattan for derailing his Steinway Commission three years earlier and who did not want the RTC to repeat his commission's mistake of substituting an elevated for an underground railway. Although suffering from gout and sometimes confined to his bed in great pain, Steinway advanced a compromise proposal for a new, scaled-down subway that had been prepared by the engineering staff in order to reduce the cost of construction below the required $50 million. It would go up the east side from city hall to Grand Central Terminal, then across Forty-second Street to Longacre Square (now Times Square) on the west side, and then up Broadway to Ninety-sixth Street, where it would divide into two branches, with one branch following Broadway through Harlem, Washington Heights, Fort George, and Riverdale in what is now the Bronx, and the other branch running up Lenox Avenue, crossing the Harlem River, and going to Bronx Park.

To be sure, Steinway's route had serious flaws. For instance, it excluded the Wall Street financial district, the fashionable Ladies' Mile retail shopping zone (located on Broadway, from Tenth to Twenty-third streets), and the Upper East Side. These omissions would subvert the original goals of easing street traffic congestion and supplying comprehensive service to all parts of the island. Yet Steinway and several other commissioners argued that these shortcomings could always be corrected in the future as long as a start was made now; half a subway was better than no subway at all. On August 6, 1896, the Rapid Transit Commission accepted Steinway's compromise. It was the old man's legacy to his adopted city; he died a few months later, on November 30, 1896.[31]

Once this new plan was accepted, the commissioners had to negotiate the review process all over again. This time around, the RTC encountered relatively few obstacles, and by March 1898, Mayor Strong, the Board of Aldermen, another citizens' review panel, and the appellate division of the supreme court had all ratified the proposition.[32]

The next step, the drawing of a contract for the subway's construction and management, took a year and a half. Finally, on November 13, 1899, the Rapid Transit Commission opened the bidding for its franchise. Two people, Andrew Onderdonk and John B. McDonald, responded. Onderdonk proposed to build the subway for $39.3 million and to share a percentage of his annual operating profits above $5 million with the municipality. McDonald submitted a flat $35 million construction bid. Not convinced that the subway would make enough money to justify Onderdonk's profit-sharing scheme, the RTC awarded the contract to low-bidder McDonald on January 16, 1900.[33]

By the time he received the subway franchise in 1900, the short, stocky McDonald had over thirty years of contracting under his belt. Born in Cork, Ireland, one year before the potato blight struck in 1845, McDonald and his family escaped to New York City in 1847, settling near Jerome Park in the Bronx. To help his struggling family make ends meet, McDonald quit school at an early age and took a job as a timekeeper on a reservoir that was being built in Putnam County as part of the Croton Aqueduct waterworks. After working on several other construction projects as a low-level supervisor, Mc-

Donald became a subcontractor for Dillon, Clyde and Company on a difficult stretch of a railroad tunnel that Cornelius Vanderbilt was digging under Park Avenue. On the strength of his success with that demanding assignment, McDonald started to win contracts for railroad projects throughout eastern North America.

McDonald knew how to manage his laborers and how to finish jobs on time and on budget. In addition, he had two special qualifications for the subway project. First, he had recently completed the first major underground electric railroad ever constructed in the Western Hemisphere, an 1894 tunnel that carried the main line of the Baltimore & Ohio Railroad under the Patapsco River in Baltimore. The lessons from this pioneering engineering work would benefit the New York subway. Second, McDonald enjoyed a warm relationship with Tammany Hall and was particularly close to grand sachem Richard Croker, who had slipped him a number of fat contracts. This tight bond between the Irish-American builder and the Irish-American politicians would allow McDonald to serve as a link between the Rapid Transit Commission and Tammany Hall and establish a quiet understanding between the two adversaries. In return for protection from political disruptions, McDonald apparently funneled jobs and subcontracts to the Tammany faithful.

In spite of his sterling engineering and political credentials, however, John B. McDonald had one serious drawback that almost put him out of the running for the subway: a lack of capital. The terms of the franchise required McDonald to deposit a $7 million bond with the city comptroller or else forfeit the contract. McDonald could not raise this huge sum on his own, so he approached August Belmont, a wealthy banker who was the head of his own investment house and the American agent for the Rothschild houses of London and Paris.[34]

Although August Belmont was not widely known as a transit investor at the time, he had long been attracted to—as he told the Rothschilds in 1892—"the splendid opportunities . . . for making a great deal of money out of schemes for improving the transportation in our large and growing cities."[35] In 1887, Belmont tried to sell the Rothschilds on an early proposal for a Manhattan subway, and three years later he served on a short-lived New York City Rapid Transit Commission. He paid close attention to the progress of mass transit in Brooklyn, Philadelphia, Chicago, and elsewhere. As a wealthy banker who understood mass transit, had the ear of the Rothschilds, and had strong Democratic party connections, Belmont was a logical choice to finance New York's subway.

On February 21, 1900, Belmont signed a contract, known as Contract No. 1, to build, equip, and operate the railway for a period of fifty years, and renewable for another twenty-five years. The municipal government agreed to give Belmont $35 million to cover the cost of construction, plus $1.5 million to buy land for the stations and terminals. In return, Belmont agreed to supply the cars, signals, and other equipment out of his own pocket. When the subway opened, Belmont would pay an annual rental equal to the interest on the construction bonds, plus a small sum for a sinking fund.

After signing this contract, Belmont created the corporation that would build the subway: the Rapid Transit Subway Construction Company; in April 1902 he formed the company that would operate the facility: Interborough Rapid Transit Company (IRT).[36]

Although John B. McDonald became the subway contractor and built the IRT, he faded into the background after Contract No. 1 was signed; August Belmont was now the man to be reckoned with.

August Belmont

August Belmont was the son of a poor but ambitious German Jew named August Schönberg who was born in the Rhineland-Palatinate in 1816. At thirteen Schönberg left his small village for bustling Frankfurt, where he landed a job with the leading Jewish banking house in Europe, the Rothschilds. Schönberg began by sweeping the floor, but he gradually proved himself to the Rothschilds to be a skilled, trustworthy money man, and they rewarded him with a number of important assignments, including a delicate negotiating mission to the Papal Court in Italy.

In 1837 the Rothschilds asked Schönberg to be their representative in Cuba, where the family had valuable sugar holdings that were at risk due to political turmoil. Traveling to Havana via New York City, Schönberg arrived in Manhattan in May 1837, one week after the onset of a financial panic that had wiped out more than 250 banks, counting houses, and insurance companies. This panic paralyzed Wall Street with fear, but Schönberg viewed the emergency as an opportunity. Forgetting all about Cuba, he established his own bank and then used his shrewd financial judgment as well as his Rothschilds references to scavenge the remains of bankrupt firms. With this bold maneuver—accomplished when he was just twenty-one years old—Schönberg made a small fortune.

Schönberg encountered a serious obstacle when he tried to infiltrate New York's commercial aristocracy: anti-Semitism. He was not its only victim; the Seligmans, Kuhns, Loebs, and Guggenheims experienced the same hostility. These German Jewish families responded by creating a parallel aristocracy that served to preserve their Jewish identity at the same time that it rewarded their material success and maintained their commercial ties with gentile firms.

But August Schönberg wanted to be a blue blood himself. Unlike his fellow German Jewish immigrants, Schönberg shed his Jewishness altogether and contrived a new identity. He began the process by changing his name. He dropped Schönberg (which means "beautiful mountain" in German) for its French equivalent, the less Jewish-sounding Belmont.[37] August Belmont adopted the religion of New York's patriciate, Episcopalianism; he enhanced his pedigree by marrying the daughter of Commodore Matthew Calbraith Perry, a national hero who had opened Japan to the western nations in 1854 and who was the brother of Oliver Hazard Perry, the victor of the Battle of Lake Erie during the War of 1812. Belmont bolstered his social standing by throwing elaborate balls at his Fifth Avenue mansion and by patronizing the Union Club, the Academy of Music, and other old-guard institutions. Thanks to his wealth, tenacity, and disingenuousness, Belmont the elder succeeded in granting his son August entry into the social elite at the very moment of his birth on February 18, 1853.[38]

Young Belmont grew up in a world of enormous privilege. He attended the Rectory School, Phillips Exeter Academy, and Harvard College before entering his father's bank in 1874. When his father died sixteen years later, August Belmont inherited the presidency of the bank and a seat at the head table of American capitalism. Belmont maintained his family's close relationship with the Rothschilds, spoke to J. P. Morgan, John D. Rockefeller, and William C. Whitney on a first-name basis, and served on dozens of corporate boards. An active Democrat who financed candidates in local, state, and national elections, Belmont helped lead the conservative "goldbugs" who fought the populist-leaning William Jennings Bryan for control of the party.

Belmont lived like a prince. He owned seven different homes, any one of which would have been the pride and joy of an ordinary millionaire.

An active sportsman all his life, Belmont used games to signify his high social position. He liked to golf, hunt, fish, and sail, but he preferred the kind of sports that combined animal breeding with personal competition among the upper crust. He loved horse racing most of all. As the most influential horseman of his day, Belmont used his powerful position as chairman of the Jockey Club to dominate the racing world by setting the racing calendar, licensing trainers and jockeys, and guarding the purity of thoroughbred bloodlines. In 1902 he bought a 650-acre tract of land on the Queens-Nassau border and set out to build the best racetrack in the country. Opened three years later and named for his father, Belmont Track boasted a 650-foot-long grandstand, a mile-and-a-half track, and an opulent clubhouse that housed a luxurious dining room, bars, and bedrooms. Belmont owned a fine stable of racehorses, too. Of all his many accomplishments, Belmont might have been proudest of having bred Man o' War, perhaps the greatest thoroughbred the American turf has ever produced. Man o' War (who was sold as a yearling and never wore Belmont's maroon and scarlet colors on the track) perfectly embodied his owner's values of good breeding and head-to-head competition.

Belmont's foul temperament was his main failing. The fat little banker was arrogant, pompous, mean-spirited, and quick to anger. He sometimes flew into a rage when not accorded the obsequious treatment he thought was due a gentleman of his elevated station. When a rude trolley car conductor offended him one day, for instance, Belmont fired off a whining complaint to the railway's president demanding redress for this grievous injury. And when the dean of Harvard College refused to excuse his son from school to recover from a cold on a ten-day convalescence in the South, he angrily dispatched a three-page letter to the college president protesting the dean's action as a stain on his honor. Generally, however, Belmont felt no contempt for his inferiors because he usually remained oblivious to their existence. Belmont felt that his own breeding set him apart from the common herd and that his triumphs in business and on the playing field verified his pedigree.

In signing Contract No. 1, August Belmont hoped to make a lot of money and to receive popular acclaim as a benefactor of the city. To his dismay Belmont would discover that mass transit was no place for a thin-skinned tycoon who could not tolerate the rough give-and-take of urban politics. Ultimately, he would regret his decision to finance the IRT, even though it earned him a fortune.[39]

3

WILLIAM BARCLAY PARSONS AND THE CONSTRUCTION OF THE IRT

The Geology of Manhattan

Manhattan's geology is forbidding. Although the island has a total of only twenty-three square miles, it harbors an unrivaled range of geological features that posed a severe challenge to the IRT's builders.

Manhattan is shaped like an irregular rectangle. Seven miles long from the Battery to the Harlem River, the island has a width of roughly two miles up to 125th Street, and then it tapers into an elongated neck that ends at Inwood. It has three main topographical zones. The first zone encompasses Manhattan's southern end, the tongue of land below 23rd Street. The terrain there is flat and featureless, rising almost imperceptibly from sea level at the Battery into low, undulating hills that have long since been covered by a grid of streets and buildings. The topography of the second zone, from 23rd to 103rd Street, is much rougher, with steep hills and rocky outcroppings that gave a rough, corrugated appearance. The natural landscape of this zone, like that of the first, has been all but erased.

The third zone, above 103rd Street, is the most varied. Northern Manhattan is dominated by a line of ridges along its western shore. The Manhattan Ridge goes from 103rd Street to about 160th Street, where it divides into two spurs: Fort Washington Ridge, which parallels the Hudson River to the head of the island, and Fort George Ridge, which borders the Harlem River as far north as Dyckman Street. These ridges increase in elevation from south to north, peaking near 185th Street and Fort Washington Avenue, where a rocky point rises 268 feet above sea level, the highest point on the island. Upper Manhattan also has two plains: the broad Harlem lowlands, which stretch east from Morningside Avenue and St. Nicholas Avenue to the Harlem River, and the smaller Inwood lowlands, above Dyckman Street on the Harlem River. In addition, two major faults bisect the island diagonally from northwest to southeast. One fault crosses from 125th Street on the Hudson River to Ninety-sixth Street on the East River, and the other follows Dyckman Street from the Hudson River to the Harlem River. The consequence of water seeping into areas of intensively shattered rock since the Paleozoic era 600 million years ago, these faults cut through the bedrock hundreds of feet below the sea. Although partially filled with glacial deposits of sand, silt, and gravel that raised the level of the ground, the two faults remained important barriers to transportation at the turn of the century.

The island's rock formations are formidable, too. Although a relatively soft rock, called dolomitic marble, overlays the Harlem and Inwood lowlands, the rest of the island is covered by a much more difficult type of rock, Manhattan schist. A well-foliated rock that has alternating black and light gray bands, schist weathers to a dark brown or black color after being exposed to the air and is speckled with coarse mica flakes that shimmer in the sunlight. Manhattan schist can thwart subterranean construction. A metamorphic rock forged deep within the earth's crust under intense pressure and great heat, Manhattan schist is a very hard rock that is murder to cut through. More serious than its hardness, however, is the fact that it is not uniformly hard. Most formations are strong and durable, but this rock is susceptible to decay and can fracture or collapse without warning. This combination of overall hardness plus occasional decay makes schist highly unpredictable and dangerous.[1]

William Barclay Parsons

The man who had to tackle this geological nightmare was William Barclay Parsons, the chief engineer of the Rapid Transit Commission. Parsons was the scion of a proud, old New York family that had belonged to the city's Anglo-American elite during the colonial era. His mother was a descendant of one of North America's most powerful families, the Livingstons of upstate Columbia County; his father was the grandson of the Reverend Henry Barclay, the second rector of prestigious Trinity Church, and a distant relative of Colonel Thomas Barclay, a Tory who fought with the British army during the American Revolution. Parsons' forebears lost most of their political clout when independence severed their ties with the mother country, but they continued to identify closely with Great Britain throughout the nineteenth century as a way of underscoring their elite social standing.

Born on April 15, 1859, William Barclay Parsons inherited a strong dose of Anglophilia from his family. At twelve his parents sent him to an English private school in Torquay, Devonshire, and for the next four years he studied under private tutors while traveling through Britain, France, Germany, and Italy. On this grand tour Parsons formed a lasting admiration for the British upper class as an aristocracy of lofty birth and solid achievement. A dyed-in-the-wool imperialist who credited the British aristocracy with raising the level of civilization around the world, Parsons always tried to emulate its high standards. He went as far as taking his English-sounding middle name, Barclay, as his personal sobriquet, retaining his more ordinary first name, William, for official purposes.

In 1875, Parsons returned to New York and enrolled at Columbia College. A big, strapping youth who was good at games and was popular with his classmates despite a humorless streak that earned him the unattractive nickname "Reverend Parsons," he stroked the heavyweight eight-oared crew to a string of victories on the Harlem River, captained the tug-of-war team, won election as class president, and cofounded the student newspaper, the *Spectator*. He graduated from Columbia College in 1879 and three years later received a degree in civil engineering from the university's School of Mines.

After a brief stint on the staff of the Erie Railroad, Parsons started his own private practice as a consulting engineer in January 1885, with offices at 22 William Street in lower Manhattan. Although he designed a number of water supply systems and railways, Parsons became particularly excited about a scheme that a group of New York City businessmen hatched to build a subway from the Battery to the Harlem River. These entrepreneurs organized the Arcade Railway Company (which, despite its name, was not related to Alfred E. Beach's pneumatic railway) and hired Parsons as a staff engineer. The Arcade was riddled with internal strife, and Parsons and several other dissident employees soon split away to create a rival enterprise, the New York District Railway. The Arcade and the District spent more time fighting each other than planning subways, however, and both companies quickly went bankrupt.

Instead of being disheartened by this failure, Parsons grew fascinated by rapid transit engineering. He pored over topographical maps of Manhattan and hiked through city neighborhoods trying to figure out the best route, motive power, and construction methods for an underground railway. At the same time he exploited his aristocratic connections by cultivating Abram S. Hewitt, William Steinway, Seth Low, and other blue-blooded subway promoters.

Parsons' lobbying paid off. In 1891 the Steinway Commission chose him as its deputy chief engineer. As the principal drafter of the Steinway Commission's subway plan, Parsons earned the respect of other engineers and drew even closer to Hewitt, Low, and Steinway. When the new Rapid Transit Commission replaced the Steinway Commission in 1894, Parsons became its chief engineer.[2]

William Barclay Parsons, thirty-five, now took direct control of subway building. Tall and rangy, with a prominent jaw and piercing eyes that made him appear good-looking in a rough sort of way, Parsons radiated strength and dignity. He stayed calm under pressure and was known for his rigid self-control. Parsons was hardly an amiable or engaging man, and he had little personal warmth. But this stern, demanding patrician nonetheless drew first-rate engineers to his side, inspired their best work, and earned their lifelong loyalty.

A journalist later observed that Parsons was a "born general and diplomat" who was "as thorough as a machine" and possessed an extraordinary gift for leadership.[3]

True to his upper-class background, Parsons had a keen sense of social responsibility and believed that engineering entailed much more than narrow technical considerations. He thought of engineering as an instrument for expanding America's wealth and power so that it would eventually surpass Great Britain as the dominant world power. Parsons was an entrepreneurial engineer who worked with capitalists such as J. P. Morgan and August Belmont. At first on his own and later as the founder of a top engineering firm that became known as Parsons Brinckerhoff, William Barclay Parsons built important public works aimed at enlarging the country's commercial domain. In addition to consulting on the Panama Canal, Parsons built docks in Cuba, the Cape Cod Canal in Massachusetts, and hydroelectric plants across the United States.

Parsons found his life's work with the subway. Fervently embracing Abram S. Hewitt's imperial vision of rapid transit as an instrument for guaranteeing New York City's future, Parsons thought of the subway as a mission rather than a mere job.[4]

The Age of Electricity

The first decision confronting William Barclay Parsons and the Rapid Transit Commission was the selection of a motive power for the subway. Their choice was influenced by a technological revolution that had swept surface transit: the adoption of electricity as a new propulsion system.

Engineers had grasped the fundamental principles of electric traction as early as the 1830s. Electricity provided a means of transferring power from the point where it was produced to the point where it was consumed. A coal-burning stationary steam engine, located at a central place, produced mechanical energy, which was converted into electrical current by a machine called a dynamo and then transmitted instantaneously via a conductor to a railway car, where a second machine, the motor, turned it back into mechanical power that drove the wheels. Electricity seemed to have limitless potential as a source of clean, fast, inexpensive, and lightweight energy that could replace the increasingly inadequate animal and steam power. For decades, however, vexing technical problems blocked the development of a satisfactory electric railway, and nobody came close to designing a commercially successful electric railway until the 1880s.[5]

The pioneer of the electric railway was Frank J. Sprague. A graduate of the U.S. Naval Academy and a former assistant to Thomas A. Edison, Sprague founded his company in 1884 to develop electrical motors for factories as well as for railways. In May 1887, Sprague signed a contract with the Union Passenger Railway Company of Richmond, Virginia, to electrify its twelve-mile route. This was a daunting task. The path of the railway wound through a hilly section of the Virginia capital that would test Sprague's new and untried system to the limit, and he also had to improvise much of his equipment on the spot. For instance, he experimented with fifty different trolley poles before finally coming up with one that maintained contact between the overhead wire and the motor.

Sprague beat the odds. Within three months of its February 1888 opening, the Union Passenger Railway was running thirty trolleys at once without experiencing any serious mechanical problems. More important, early electric railways such as Sprague's Union Passenger Railway performed better than competing transportation modes. Their capital costs were 80 percent lower than those of cable railways; their operating costs were 40 percent lower than those of horse railways; and their average speed of twelve miles per hour was about three times faster than that of horse railways.

Frank J. Sprague proved that electrification could be commercially successful on a large scale and under adverse conditions. Consequently, electric street railway construction surged. When Sprague received his Richmond contract in 1887, there were only eight street railways using electric power in the United States (with a mere 35 miles of track). Fifteen years later, in 1902, the total length of electric surface track had multiplied to 21,920 miles. By then the trolley had completely replaced the horse railway as the chief form of surface transport.[6]

But Sprague's Richmond triumph did not ensure that the New

York subway could be electrified. When William Barclay Parsons became chief engineer of the Rapid Transit Commission in 1894, electrical traction was still in its infancy in the United States. Only a few rapid transit lines had been electrified so far, such as Chicago's Intramural Railway, a temporary demonstration line built for the World's Columbian Exposition of 1893. Thus, Parsons could not be confident that a technology devised for a twelve-mile-long street railway in a sleepy southern town could be adapted to a high-capacity, high-speed subway for North America's largest metropolis. He remained skeptical about electric traction's possibilities and prudently refused to rule out steam or cable.[7]

Late in the summer of 1894, Parsons sailed for Europe to make a survey of its railways. Because Europe, not North America, was in the forefront of rapid transit operations, it was the best place for Parsons to pick up the technical knowledge he needed to make the right decision. He traveled to Glasgow, where a cable-powered underground railway was being dug that in two or three years would make the Scottish municipality the second city in the world to have a subway; to Paris, where a steam-driven railway circled the urban perimeter; and to Liverpool, where an electric elevated had recently opened along the Mersey.[8]

Parsons went to London, too. London was the birthplace of underground mass transportation, and in the 1890s it remained the center of rapid transit technology. On November 4, 1890, the City and South London Railway had inaugurated the world's first electrically powered subway. Originally intended for cable power, the City and South London converted to electricity at the last moment, and its design resembled a toy train set more than a major urban railway. The new underground was only three miles long, going from the Monument in the city, to Stockwell on the south bank of the Thames River, and its two tunnels had a diameter of just ten feet two inches. Its four-car trains were tiny, seating no more than ninety-six riders, and they averaged just thirteen miles per hour. The passenger accommodations were not ideal, either. Angry riders complained that the small, virtually windowless carriages were stuffy and uncomfortable and that the movement of the trains through the iron-ribbed tunnels created an irritating ringing sound.

Despite these drawbacks, Parsons understood that the City and

South London represented a gigantic step forward from London's two steam undergrounds, the Metropolitan Railway and the Metropolitan District Railway. He noted that its stations and carriages were clean and well ventilated, and that its patrons did not ruin their clothes or inhale noxious fumes. But Parsons was more concerned about economics than about the environment. Consequently, he was impressed that the City and South London tallied lower fuel, repair, and labor costs than the two steam undergrounds, suffered relatively few mechanical breakdowns, and made a modest profit. To Parsons this was convincing proof that electricity was the solution.

After returning to New York, Parsons wrote a report for the Rapid Transit Commission advocating the electrification of the subway. Parsons' brief, entitled "Report on Rapid Transit in Foreign Cities," removed the commissioners' last doubts about this new energy source and brought the lengthy search for a motive power to an end.[9]

How Deep Should the Subway Go?

As much as he had learned from the City and South London, Parsons knew that the New York subway could not be designed from another underground railway's blueprint. Manhattan's tough physical and urban geography demanded a unique engineering solution, not a carbon copy of another's.

Manhattan's geological uniqueness had a particularly strong bearing on the question of the subway's depth. The world's first underground, London's celebrated Metropolitan Railway, had been constructed near the surface in order to vent the ashes, cinders, and smoke that its coke-burning locomotives spat out. But the use of electricity eliminated these pollutants from the tunnels, allowing the City and South London to be buried an average of fifty feet below the ground. The City and South London passed mostly through a type of soil called London clay, which was highly stable and uniform. Dense, cohesive, impervious to water yet relatively soft, London clay was perfect for tunneling. Consequently, most of London's subsequent tubes were also built as deep tubes.[10]

New York City was another case. Geologists and engineers recognized that Manhattan schist was a hard, heterogeneous substance that would impede tunneling. What they had not appreciated, however, was that the schist did not lie at an even depth below the surface. In the nineteenth century the amount of knowledge about New York City's geology was minimal. Before the construction of skyscrapers required digging deep foundations, and only a few public works such as the Croton Aqueduct and the Brooklyn Bridge entailed much underground construction, engineers lacked basic information about the subsurface. William Barclay Parsons himself was among the first to discover that the distance from the ground to bedrock varied from place to place. In 1891, while drilling test holes for the Steinway Commission along the route of its projected Broadway subway, Parsons uncovered a striking phenomenon: The schist came within 20 feet of the surface at Whitehall Street on the Battery; dropped to a depth of 163 feet at Duane Street, three blocks north of city hall; remained at a low level under Greenwich Village; and then climbed back to 16 feet at Thirtieth Street. This U-shaped rock contour had important consequences for Manhattan's growth. For instance, one reason for the emergence of lower and midtown Manhattan as the main business districts was that the schist rose so close to the surface there, it provided an excellent building foundation for skyscrapers.[11]

Parsons' discovery affected the decision of the Rapid Transit Commission about whether to construct the New York subway near the surface, like the Metropolitan, or far below ground, like the City and South London. If the RTC followed the City and South London's lead, its tunnel would cross a mixed face of partly soft earth and partly hard rock. Transiting a mixed face was a poor engineering practice because it would generate high construction costs and produce an unsound structure. One way to avoid a mixed face was to sink the tunnel so deep—as much as two hundred feet below Broadway, the equivalent of two-thirds of a football field—that it would pass only through solid bedrock. But even though this alternative would bring about a structurally sound subway, Parsons warned that it would require the installation of expensive elevators, escalators, and ventilators that would raise capital and operating costs to uneconomical heights.

The answer, Parsons argued, was to build the subway within

fifteen or twenty feet of the surface. He claimed that a shallow subway could be constructed by excavating a trench in the street, a relatively simple expedient that would cost less than a deep tube and yet would yield a stable structure. There was one drawback to a shallow subway, however: The space below most city streets was already filled with a maze of electric cables, telephone lines, telegraph wires, water pipes, steam mains, and sewers that would have to be removed and rebuilt elsewhere. Parsons nonetheless concluded that a shallow subway would cost one-eighth as much as a tube, and he strongly recommended that the Rapid Transit Commission opt for a shallow subway.[12]

In early 1895 the Rapid Transit Commission adopted an engineering plan that incorporated most of Parsons' ideas: a shallow railway with four tracks on a single level as far as the Ninety-sixth Street junction, and with two tracks (later enlarged to three in some places) on the Broadway and Lenox Avenue branches.[13] Once the RTC approved its basic engineering design, Parsons had to wait until the commissioners awarded a franchise and then construction could begin. For the chief engineer, the five-year wait for the RTC's plan to emerge from the labyrinth of mayoral, aldermanic, and judicial oversight was torture. Parsons grew particularly concerned when other European and North American cities began passing New York City in the race for rapid transit. Between 1895 and 1900, as New York's courts dickered over the RTC's budget and route, Glasgow, Budapest, and Boston unveiled new underground railways and Paris started to build its first metro. Parsons feared his dream of building the New York subway would not be realized.[14]

Construction of the IRT

Late in 1899 a cable announcing that the RTC's subway plan had finally been approved reached Parsons in Canton, China, where he was surveying a thousand-mile railroad from Hankow to the sea on behalf of a syndicate headed by J. P. Morgan. Overjoyed with the happy news, Parsons immediately quit the Chinese survey and sailed for home. By the time subway construction began on March 26,

1900, two days after a groundbreaking ceremony was held on the steps of city hall, Parsons was already hard at work.

Contemporary building technology was so primitive that the IRT had to be constructed almost entirely by hand. Because there were few steam shovels or bulldozers available in 1900, the burden fell almost entirely on the seventy-seven hundred laborers who made up the workforce at its peak. These workers were predominantly Irish and Italian, although there were also Germans, African-Americans, Greeks, and members of other ethnic groups. Unskilled workers earned from $2.00 to $2.25 per/day, skilled workers about $2.50. Wielding picks, shovels, hammers, percussion drills, and other hand tools, these workmen literally gouged the subway out of the raw earth.[15]

These laborers led hard lives. For instance, one worker who sustained a serious injury building the IRT was an Italian immigrant named Salvatore Mazzella. Born on May 6, 1883, on the small island of Ponza off Naples, Mazzella emigrated to New York City at age seven with his father and grew to be a handsome, blue-eyed six footer. Trained as a tile worker, Mazzella landed a job cementing white tiles to the walls of the IRT stations. He apparently labored on the subway for five or six years, first on the original IRT and then on its extensions; typically, he worked on the IRT during the summer and then returned with his savings to Ponza for the winter. At some point in 1908 or 1909, Mazzella spilled some lead-based tile adhesive in his eye. Mainly because he spoke English poorly, Mazzella did not tell his Irish foreman about the accident until after his eye became badly inflamed. His boss eventually found out and urged him to see a doctor, but it was too late: The doctor had to remove his eye and provide him with a glass replacement. Salvatore Mazzella remained proud of his contributions to New York City for the rest of his life, but he paid a high price for that sense of accomplishment.[16]

Thousands of workers like Mazzella built the IRT subway piece by piece. The standard form of construction was known as cut and cover. The laborers began by cutting a hole the width of the street. Excavating this trench was fairly easy below Tenth Street, where the subway ran through soft soils. It was much more difficult above Tenth Street where much of the route passed through solid rock that had to be drilled and dynamited. In both sections, work gangs had to shore up the old buildings that lined the path of the IRT, maneuver around the underground storage vaults that extended into the streets, and relocate sewers and utilities. Traffic was a concern, too. On busy downtown thoroughfares such as Park Row that had to stay open during construction, workers covered the trench with a temporary wooden bridge that supported trolleys, carriages, and wagons; on less important arteries the street was closed during construction and the hole remained open.

After completing the trench, the workers erected the structural framework that would house the tracks, signals, third rail, and other equipment. This framework was made of steel and concrete and resembled an elongated rectangular box. Laborers poured a four-inch-thick concrete foundation across the bottom of the trench to form the box's base. They fabricated its sides and roof by planting steel I-beams every five feet and pouring concrete into the gaps between the beams. This technique of embedding steel columns in concrete produced an exceptionally strong structure that could easily bear the weight of the street.[17]

Cut and cover was the most common type of construction, but it could be employed on no more than 52 percent of the subway's total length. Due to the island's hilly topography, abrupt changes in the ground level occurred so frequently that the use of cut and cover would not have kept the rails at grade. To prevent the IRT from resembling a Coney Island roller coaster, the RTC had to build a wide variety of structures, including a 2,174-foot steel arch viaduct across Manhattan Valley between 122nd and 135th streets and rock tunnels in Murray Hill and upper Manhattan.

Of all the techniques used on the subway, rock tunneling was the most demanding. The workers started by sinking a vertical shaft at both ends of the tunnel. Then they isolated a small section at the bottom of each shaft and began driving a narrow heading there. This heading was advanced by drilling holes seven feet into the face of the schist and putting dynamite charges in the cavities. After detonating the explosives, the laborers returned to the face, cleared the rubble, and braced the new section of the tunnel. Then they started drilling again. This cycle of drilling, blasting, clearing, and timbering continued until the two headings met in the middle and the tunnel was completed. The workmen then enlarged the tube to its full size, lined it with concrete, and installed tracks, third rails, and signals.[18]

Boring through Manhattan schist was difficult because the rock was hard; it was dangerous because the schist lacked uniformity. Drillers often encountered unexpected geological hazards that delayed the project and cost lives—rock slides, pockets of broken rock or loose gravel, and underground ponds. The Murray Hill tunnel, which went up Park Avenue from Thirty-fourth Street to Forty-first Street, illustrated these dangers. A rocky promontory that rose thirty or forty feet above the surrounding landscape, Murray Hill covered the area from Twenty-seventh Street to Forty-Second Street, and from Third Avenue to Sixth Avenue. Although it was not as prominent geologically as the ridges of upper Manhattan, Murray Hill consisted of a highly unstable formation of schist that posed great obstacles to subway building. The rock lay at a forty-five-degree angle that could precipitate rock slides, and it harbored many belts of shattered rock that could lead to cave-ins. The construction of the IRT's Murray Hill tunnel was further complicated because an old trolley tunnel of the Metropolitan Street Railway Company ran below Park Avenue, right in the path of the subway. Consequently, the subway had to be divided into two separate tunnels and shoehorned into the remaining space.

The subcontractor for this section of the subway, thirty-eight-year-old Major Ira A. Shaler, experienced misfortune after misfortune. His first calamity occurred on January 28, 1902, when a wooden powder house, located at Forty-first Street and Park Avenue and containing over two hundred pounds of dynamite, caught fire and exploded, giving off a blinding white flash, a column of dirty yellow smoke, and a powerful shock wave. One New Yorker who witnessed this devastation recalled his fright: "I thought the end of all of us had come."[19] The blast wrecked the Murray Hill Hotel, defaced Grand Central Terminal, and shattered glass for several blocks around. Five people died and another 125 were injured. A grand jury blamed the accident on Shaler's carelessness in storing the explosives and indicted him for manslaughter.

Disaster struck again two months later, on the night of March

21, when a narrow thirty-five-foot crack appeared in the roof of the east tunnel, below Thirty-seventh Street. In trying to shore up the tunnel, Shaler's employees made the alarming discovery that the roof was not covered by a solid mass of schist, as it should have been. Instead, the roof consisted of only a thin shell of hard rock. Above this shell was a loose heap of decomposed rocks, pressing down with the full force of gravity. At 8:30 the next morning an avalanche of boulders, stones, and dirt crashed through the roof with a deafening roar. Although the workers had been evacuated and nobody was hurt, debris partially filled the tunnel, delaying subway construction for weeks. In addition, the break extended to the surface of Murray Hill and undermined the foundations of four homes, causing over $100,000 in property damage.

By now Shaler had been dubbed the "voodoo contractor." His luck ran out completely three months later, on June 17, 1902, when he took Chief Engineer Parsons and Assistant Engineer George S. Rice on an inspection tour of his section. As they reached the upper end of the western tunnel, Parsons stopped, pointed his cane at a rock protruding overhead, and warned that it looked rotten. Major Shaler, insisting that Parsons was wrong and that the rock was perfectly safe, stepped out from the timber bracing for a closer look at the ledge. Suddenly, a one-thousand-pound boulder collapsed. Parsons and Rice had remained under the timber and were not even scratched, but the rock broke Shaler's neck and paralyzed him below the shoulders. He died in Presbyterian Hospital eleven days later.[20]

The subway's most daunting structure was a tunnel that ran over two miles from 158th Street in Washington Heights to Hillside Avenue in Fort George. It ranked as the second longest two-track rock tunnel ever built in the United States, surpassed only by the famous Hoosac Tunnel in western Massachusetts. It was so deep that elevators had to be installed so that passengers could reach the 168th Street, 181st Street, and 191st Street stations; the 191st Street stop is 180 feet below ground and remains the subway system's deepest station today.

The construction of this imposing tunnel attracted miners from all over: eastern Pennsylvania's anthracite coal hills, Colorado's silver lode, the Klondike gold strike, South Africa's gold and diamond fields, Wales, Ireland, Scandinavia, and Canada. These were miners who spent their lives wandering from place to place. They earned up to $3.75 a day, almost twice as much as unskilled workers, and pointedly referred to the tunnel as the "mine" in order to distance themselves from ordinary laborers. During the IRT's construction they lived in boardinghouses in Washington Heights and turned this neighborhood into a tough mining camp that reverberated with the rhythms of work, drinking, and whoring.[21]

The single worst accident that took place during the construction of the subway occurred here in October 1903, near the tunnel's Fort George portal. At the time the tunnel crews were working several hundred feet north of 193rd Street and St. Nicholas Avenue, on the rear slope of Fort George, where the schist lay at a steep angle that was conducive to slides, and in a strata of broken rock that was riddled with fissures and seams. Although these treacherous geological conditions probably should have dictated a cautious approach, only a few hundred feet remained to be excavated before the tunnel was holed through, and the contractor, L. B. McCabe & Sons, was in a hurry. Consequently, McCabe ordered three dynamite blasts be set off every day instead of the normal two.

This decision proved fatal. Shortly after 10:00 P.M. on October 24 a gang of twenty-two men triggered a series of explosions in the tunnel. Ten or fifteen minutes later foreman Timothy Sullivan returned to the rock face by himself in order to sound the tunnel's walls. He was responsible for making sure that the blast did not loosen any rocks in the roof and walls that might fall on his crew. Thinking the schist looked stable, Sullivan shouted, "Come on, boys, let's get to work." Sullivan was unaware that an underground spring was hidden behind the face, weakening the rock. Shortly after the laborers reached the blast site, a huge three-hundred-ton boulder, measuring forty-four feet long and four or five feet wide, dropped from the roof, killing six men instantly and seriously injuring eight others. The tunnel's walls were splattered with a gory mixture of human blood and spring water, and the air was filled with the moans of badly hurt workers who were buried under mounds of timber, rock, and dirt. Although rescuers freed several of the trapped men, three Italian laborers were crushed so badly that they were beyond saving. The doctors could do no more than ease their horrible suffering with morphine injections. Meanwhile, Father Thomas F. Lynch of St. Elizabeth's Church administered the last rites. Bravely ignoring the possibility that another cave-in might happen at any moment, Lynch stayed with the three Italians until they died. The doctors did manage to liberate another man who was pinned under the boulder by amputating his leg, but he died not long after being rushed to Lebanon Hospital. Ten people died that grisly October night: Timothy Sullivan, the Irish foreman; William Schuette, a German electrician; and eight unknown Italian laborers.[22]

Two days later William Barclay Parsons arrived at the portal and inspected the accident site. Dismissing the possibility that the contractor's speedup might have contributed to the disaster, Parsons concluded that the mishap was an unavoidable result of Fort George's unforgiving geology. "All possible [safety] precautions had been taken," he assured the Rapid Transit Commission in his annual report.[23]

Parsons lost no sleep over the deaths of these lowly workers of foreign stock. Indeed, although his official report described this loss of life as regrettable, in his private diary Parsons concentrated on the geology of Fort George and did not bother to mention that anyone had died there at all.[24]

As a veteran civil engineer who knew the dangers of heavy construction, William Barclay Parsons unblinkingly accepted the ten Fort George deaths—along with those of the other forty-four workers and civilians who perished during the subway's construction—as the price that had to be paid for progress.[25] To Parsons' cold way of thinking, all that really mattered was keeping the subway on schedule and ensuring that it was a success on opening day.

He got the job done, but at a terrible price.

4

THE SUBWAY AND THE CITY

October 27, 1904

Mayor George B. McClellan switched on the motor, turned the controller, and headed north from the city hall station for a tour of the new Interborough Rapid Transit subway. The underground railway had been dedicated earlier that day in October 1904 during a ceremony at city hall, and it would be opened to the general public four or five hours later; people were already waiting outside the stations for their first ride on the new subterranean wonder. For now, however, Mayor McClellan and a select group of the city's most distinguished leaders had it to themselves.[1]

Although Mayor McClellan was supposed to surrender the controls to a regular motorman once the excursion got under way, he was enjoying himself so much that he adamantly refused to step aside and insisted on running the train himself, despite the fact that, as he admitted with a boyish grin, he had never driven a railway car before.

McClellan's decision startled the two top Interborough executives on board the train, Vice-President E. M. Bryan and General Manager Frank M. Hedley. Although they naturally wanted to indulge a politician as powerful as the chief executive of New York

City, Bryan and Hedley feared that McClellan might cause a serious accident which would turn the public against the IRT. Bryan and Hedley were concerned because the Interborough was so novel that city residents might not accept it. Although rapid transit engineering was quite sophisticated, there were just six other subways in the world by 1904, and the only other one in the Western Hemisphere was a short Boston line that had been in service for a mere seven years. Indeed, one wag had predicted years earlier that a New York subway was bound to flop because people would go below ground only once in their lifetimes—and that was after death.

Bryan and Hedley were particularly worried about the possibility of an accident because many prominent individuals were riding this special excursion train: President Nicholas Murray Butler of Columbia University and Chancellor Henry Mitchell McCracken of New York University; Archbishop John M. Farley of the Catholic Church and Bishop Coadjutor David H. Greer of the Episcopal Church; and bankers Jacob Schiff, Morris K. Jesup, and Henry Clews. These men represented elite groups that ruled New York City, and their participation in these opening day ceremonies underscored the importance of the subway. To these leaders the IRT belonged in the tradition of grand civic projects that made New York the greatest metropolis in North America, such as the Erie Canal, the Croton Aqueduct water system, and the Brooklyn Bridge.[2]

But George B. McClellan was having the time of his life driving the train, and he did not share the IRT executives' anxieties. Mischievously ignoring the pained expressions on the faces of Bryan and Hedley, McClellan pressed the controller and sent the special train shooting through the tunnel. General Manager Hedley nervously glued himself to the emergency brake and kept blowing the whistle to alert track workers to the danger, while Vice-President Bryan stationed himself directly behind the mayor, ready to take charge in a crisis. At a point near Spring Street, Hedley tried to coax McClellan into ending his joyride by asking, "Aren't you tired of it? Don't you want the motorman to take it?" "No, sir!" McClellan replied firmly. "I'm running this train!"[3]

So he was. One mishap did occur when McClellan accidentally hit the emergency brake, bringing the four-car train to a violent stop that sent the silk-hatted, frock-coated dignitaries flying through the air. But McClellan confidently settled back at the controls after a few tense moments and, despite Hedley's panicked cries—"Slower here, slower! Easy!"—raised the train's speed even higher.[4] Up Elm Street (now Lafayette Street) and Fourth Avenue (now Park Avenue) to Grand Central Terminal, across Forty-second Street to Times Square on the West Side, and then up Broadway through the Upper West Side, the train roared on the straightaways and careened around the curves at over forty miles per hour. Afterward the "motorman mayor," as he was now called, attributed his success to his mastery of "automobiling."

Mayor George B. McClellan's escapade was hardly the only reason this trip was memorable, for the IRT subway ranked as the world's best rapid transit railway. A British transit expert who inspected the Interborough in 1904 concluded that it "must be considered one of the great engineering achievements of the age" and acknowledged that it was more advanced technologically than London's underground.[5] The IRT was not yet the largest subway in the world—London still held that distinction—but it was the longest ever completed at a single time, covering twenty-two route miles. Because the northern sections were still under construction, the special reached only as far uptown as 145th Street and Broadway on its run. (See Map 1, Interborough Rapid Transit Company, Contract No. 1 and Contract No. 2 subways, 1908.)

President Butler, Chancellor McCracken, Archbishop Farley, and Jacob Schiff were dazzled by their first look at the IRT. The subway had been created not merely as a pedestrian municipal service but as a civic monument. The architects were Heins and LaFarge, who had designed America's largest church, the Cathedral of St. John the Divine, on New York's Morningside Heights, and made some of the subway stations works of art in themselves. For example, the flagship station at city hall was an underground chapel in the round that had beautiful Guastivino arches, leaded glass skylights, and chandeliers. Although the other stations were utilitarian boxes that could not compare to city hall's, they nonetheless had colorful tile mosaics, natural vault lighting, and oak ticket booths with bronze fittings that created an elegant impression. Their most popular feature was ceramic bas-relief name panels that depicted neighborhood themes: The Astor Place plaque had a beaver in honor of fur trader

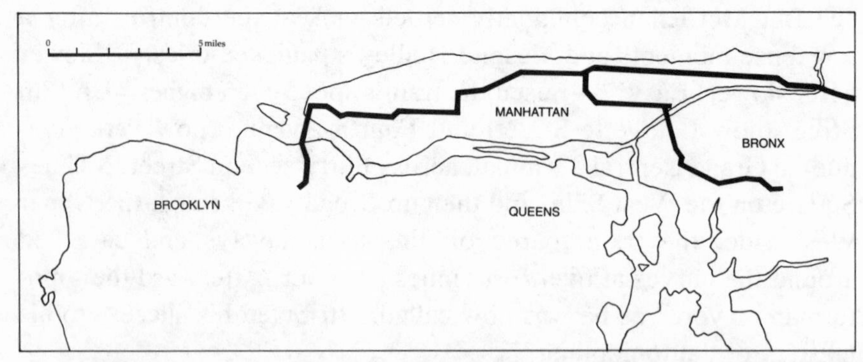

1. Interborough Rapid Transit Company, Contract No. 1 and No. 2 subways, 1908.

John Jacob Astor; Fulton Street showed Robert Fulton's pioneering steamboat, the *Clermont;* Columbus Circle had Christopher Columbus' flagship, the *Santa Maria;* and 116th Street (Broadway branch) displayed the Columbia University seal.

The Interborough was equally attractive above the surface. For instance, McKim, Mead & White had turned the facade of the IRT powerhouse—a colossal brick and terra-cotta structure on the Hudson River at Fifty-ninth Street where coal-burning furnaces generated electricity for the subway—into a classical temple that paid homage to modern industry. Still more impressive were the cast-iron and glass kiosks that covered the station entrances and exits. These ornate kiosks, which were patterned after similar structures on the Budapest metro, soon became an Interborough trademark, and for good reason. Cleverly designed to enhance rider comfort, the kiosks marked the entrances to the stations so that patrons could find the subways without getting lost on the street; they also shielded the passageways so that people were not drenched in rainstorms and segregated incoming from outgoing pedestrians so that traffic would flow more smoothly and riders going downstairs did not have to fight those going up.[6]

After the initial V.I.P. activities, the celebration took on a democratic air. In the words of a newspaperman who observed the festivities, New York City went "subway mad" over the IRT's inauguration. For the preceding few weeks, hundreds of New Yorkers had held "subway parties" to celebrate the big event. On October 27 courthouses, office buildings, shops, and private homes were decked out with flags and bunting just as on the Fourth of July; church bells, guns, sirens, and horns resounded all day long. Thousands of people began to gather around city hall early in the morning, waiting to see the dedication ceremony that would be held upstairs in the aldermanic chamber. Thousands more queued up at the kiosks, waiting for their first subway ride.[7]

The only place where the official train ran in full public view was on the viaduct over Manhattan Valley, from 122nd to 135th streets. A huge crowd of spectators went there for a look, blanketing rooftops, vacant lots, street corners, and fire escapes for blocks around. When the special emerged from the tunnel and started across the trestle, these onlookers began to cheer. The train slowed down and blew its whistle in response; sirens from local factories and from ships on the Hudson River let loose, too. The noise continued until the train reentered the tunnel and disappeared from sight.

The rest of the trip was uneventful, largely because Hedley and Bryan had finally persuaded Mayor McClellan to relinquish the controls. Very proud of himself, McClellan puffed away at a cigar and bantered with the other guests all the way back to city hall.

Finally, at 7:00 P.M., the IRT opened its doors to the public. Men and women who had been waiting all afternoon for this moment streamed down the stairways and onto the cars. A new era had begun in the history of New York City.[8]

More than 110,000 people swarmed through subway gates that evening and saw the stations and platforms for themselves. New Yorkers were so excited by their discovery of the IRT that they coined a phrase to describe the experience: "doing the subway."

The night took on a carnival atmosphere, like New Year's Eve. Many couples celebrated in style by putting on their best clothes, going out to dinner, and then taking their first subway ride together. Some people spent the entire evening on the trains, going back and forth from 145th Street to city hall for hours. Reveling in the sheer novelty of the underground, these riders wanted to soak up its unfamiliar sights and sensations for as long as possible. In a few instances high-spirited boys and girls took over part of a car and began singing songs, flirting, and fooling around. The sheer exuberance of opening night proved to be too much for others; although they bought their green IRT tickets and entered the stations like everyone else, these timid passengers were so overwhelmed by their new surroundings that they did not even attempt to board a train. All they could do was stand on the platform and gawk.

This popular hoopla climaxed three days later, on Sunday, October 30. Most New Yorkers still worked six days a week and had only Sundays to themselves. On this particular Sunday, almost one million people chose to go subway riding. The IRT was like a magnet, attracting groups from the outskirts of Brooklyn and Queens, two or three hours away. Unfortunately, the IRT could accommodate only 350,000 a day, and many people had to be turned away. The lines to enter the 145th Street station stretched for two blocks, and people grew so frustrated that police reserves had to be summoned to break up fights and restore order.[9]

That same week a public controversy erupted when the Interborough Rapid Transit Company suddenly began to install advertising placards in the stations. Although the Rapid Transit Commission had evidently always intended to permit advertising, it withheld this information from the public. Even the subway's architects, Heins and LaFarge, did not know about it. So when workers for the Interborough's advertising firm, Ward & Gow, began driving nails into the ornamental tilework and covering the walls with large, tin-framed signs, New Yorkers reacted with surprise and anger. Many thought these unsightly billboards—for products such as Baker's Cocoa, Evans Ale (''Live the Simple Life''), Coke Dandruff Cure, and Hunyadi János (''A Positive Cure for Constipation'')—detracted from the subway's stature as a noble civic monument. The *Real Estate Record and Builders Guide* blasted the placards as ''an outrage'' that ''mar the appearance of an appropriate and admirable piece of interior decoration,'' while Rapid Transit Commissioner Charles Stewart Smith called them ''cheap and nasty.''[10] Both the Architectural League and the Municipal Art Society condemned the ads and demanded their removal. Aroused by this strong negative reaction, the municipal government filed a lawsuit against the Interborough to eliminate the ads. The Interborough expected to earn

An advertisement for omnibuses manufactured by the John Stephenson Company of New York City, c. 1855. Introduced in the late 1820s, the omnibus became the first mass transit vehicle in New York City. (New-York Historical Society)

A horse car heading east on Forty-second Street, toward Fifth Avenue, c. 1889. The Sixth Avenue elevated railway can be seen in the background, at left. (NYHS)

Top: *A view of the interior of the Beach Pneumatic Railway, under Broadway between Murray and Warren Streets, showing the tunnel, car, and station, c. 1870. (Museum of the City of New York)*

Above: *Construction of the Ninth Avenue elevated railway, looking south along Ninth (later Columbus) Avenue, between West Ninetieth and West Eighty-fourth Streets, 1879. Built through an undeveloped section of Manhattan's Upper West Side, the completion of the Ninth Avenue elevated spurred residential and commercial building along this corridor. (NYHS)*

Top: *A steam locomotive pulling a train running on the Third Avenue elevated railway in Manhattan, c. 1885. The "Els" were the first form of rapid transit in New York City. Capable of speeds up to 12 m.p.h., they led to the development of new neighborhoods in Manhattan and in Brooklyn. (New York Transit Museum)*

Above: *Steam-powered trains negotiating the serpentine curve on the Ninth Avenue elevated route at 110th Street and Eighth Avenue (later Central Park West), c. 1889. This serpentine curve, which rose sixty feet above the ground, was a famous local landmark. (NYHS)*

Right: *Abram S. Hewitt. A rich iron manufacturer who served as mayor of New York City from 1887 to 1889, Hewitt made a visionary proposal in January 1888 for the creation of an advanced rapid transit railway. Hewitt's proposal helped establish a political and financial framework for the city's first subway, the Interborough Rapid Transit Company's Contract No. 1 line. (NYHS)*

Below: *William Barclay Parsons. As chief engineer of the Board of Rapid Transit Railroad Commissioners, Parsons supervised the construction of the Contract No. 1 subway. (NYHS)*

Below right: *August Belmont, with the reins, driving a coach through the gates of his mansion at Newport, Rhode Island, 1912. A wealthy banker and a leading figure in the Democratic party, Belmont was the chairman of the Interborough Rapid Transit Company, which operated New York's first subway. (MCNY)*

Above: *Cut-and-cover construction of the Interborough Rapid Transit Company's Contract No. 1 subway, in front of the Astor Library, at Great Jones Street and Elm (now Lafayette) Street, in lower Manhattan, August 15, 1901. (MCNY)*

Left: *Construction of the IRT Contract No. 1 subway, Sixty-sixth Street and Broadway, November 28, 1902. (NYHS)*

Left below: *Laborers building the Fort George tunnel on the IRT Contract No. 1 subway, in northern Manhattan, April 26, 1904. (NYHS)*

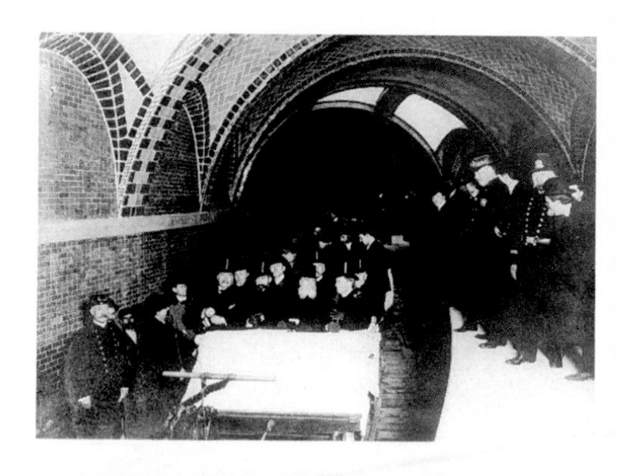

Top right: *Rapid Transit Commissioners and guests inspecting the new subway, at the City Hall station, Manhattan, February 1904. (NYHS)*

Middle right: *Mayor George B. McClellan, Jr., center, with two Interborough Rapid Transit Company executives, in the cab of the first IRT train, October 27, 1904. McClellan operated this special train for part of its inaugural run through the new IRT line. (NYHS)*

Below: *Crowds in City Hall Park celebrating the opening of the Interborough subway, October 27, 1904. (NYHS)*

A view of a subway kiosk on Broadway at 103rd Street, on the Upper West Side of Manhattan, July 30, 1913. The cast-iron and glass kiosks covered the entrances and exits to the subway stations. They became an IRT trademark. (NYHS)

Construction of the IRT Contract No. 1 subway, in Longacre Square, at the intersection of Seventh Avenue and Forty-fourth Street, Manhattan, June 4, 1901. With the opening of the IRT subway in October 1904, this area—which had been renamed Times Square several months earlier—emerged as New York's premier entertainment district. On December 31, 1904, New York Times *publisher Adolf Ochs opened his Times Building on the site of the old Pabst Building. (NYHS)*

Looking north toward Sherman Square, at the intersection of Broadway, Amsterdam Avenue, and West Seventy-second Street, on the Upper West Side of Manhattan, c. 1910. The neo-Dutch colonial IRT control house is visible in the center-right; it was designed by the architectural firm of Heins and Lafarge. (NYHS)

Top: *Panorama of the Dyckman Street station, which was still under construction, in the Inwood section of northern Manhattan, December 22, 1905. By giving easy access to outlying areas such as Inwood, the IRT subway stimulated residential construction in northern Manhattan and the Bronx. (NYTM)*

Above: *A Bronx streets department crew laying ribbons in a field in the borough's Soundview section in 1910, marking the location of the streets and avenues that were to be built there. (MCNY)*

The Brooklyn Bridge, looking west toward Manhattan, 1898. Heavily traveled by cable cars, steam-powered El trains, electric streetcars, horse-drawn wagons and pedestrians, the Brooklyn Bridge was the only bridge across the East River in 1898. (NYHS)

Rivington Street, looking west from Ludlow Street, on Manhattan's predominantly Jewish Lower East Side, c. 1910. In 1910, the Lower East Side was New York's most crowded district, with population densities exceeding 700 people per acre. (NYHS)

Right: *George McAneny, in 1909. As president of the borough of Manhattan, McAneny pressed for the adoption of the dual contracts of 1913, which doubled the length of the city's rapid transit system and eliminated the East River as a travel barrier. (MCNY)*

Below: *Construction of the Steinway tunnel extension between the Vernon-Jackson Avenue and the Hunter's Point Avenue stations, in Queens, December 10, 1915. The Steinway tunnel went under the East River between Forty-second Street, in Manhattan, and Vernon-Jackson Avenue, in Long Island City, Queens. Completed as part of the dual system, this tunnel became the first subway to link Manhattan and Queens when it opened on June 22, 1915. The extension seen here entered service in February 1916. (Vincent F. Seyfried Collection)*

Facing page top: *Barclay's Pond, on the west side of Trains Meadow Road, in Trains Meadow, Queens, c. 1900. In 1900, Trains Meadow was a rural landscape of fields, streams, and ponds. The opening of the dual system's Queensboro line in 1917 stimulated the development of this area. Today this site is located between Seventy-third and Seventy-fourth Streets and between Thirty-fourth and Thirty-fifth Avenues in Jackson Heights. (VFS)*

Facing page bottom: *A snipe hunter in Trains Meadow, Queens, c. 1900. Today, this site is located approximately at Eightieth Street and Thirty-fourth Avenue in Jackson Heights. (VFS)*

Right: *A 1923 advertisement for the Queensboro Corporation's garden suburb in Jackson Heights, Queens, emphasizing its easy accessibility to Manhattan via subway. The ad's assurance that prospective tenant-owners would be screened "for those whose social and business references are acceptable" meant that the Queensboro Corporation excluded Jews. (VFS)*

Below: *The Chateau garden apartments, Eightieth to Eighty-first Streets and Thirty-fourth to Thirty-fifth Avenues, Jackson Heights, Queens, May 1929. (VFS)*

Facing page: *The dual system subways overcame New York City's river barriers and connected four of the city's five boroughs. These new subway lines carried New Yorkers both to the Coney Island beaches in Brooklyn (above, in 1922) and to Yankee Stadium in the Bronx (below, c. 1923). (Above, Collection of the Municipal Archives of the City of New York; below, NYTM)*

ORIOLE BATHS
RESTAURANT

CAMELS
CAMELS

Mayor John F. Hylan. As mayor of New York from 1918 to 1925, Hylan defended the five-cent fare and promoted the construction of the Independent Subway System. (MCNY)

Construction of the Independent Subway System's Fulton Street line, in Brooklyn, August 17, 1931. This IND subway was built directly below the Brooklyn-Manhattan Transit Corporation's Fulton Street elevated railway, which remained in service. (NYTM)

Isac Friedlander, 3 A.M., *1939. Friedlander's etching shows ten passengers on the IRT's Lexington Avenue local. Despite this late hour, none of the riders seems to be worried about safety. The absence of fear of crime was essential to the subway's broad popularity in the 1920s and 1930s. (MCNY)*

Fritz Eichenberg, Subway, *1934. This 1934 woodcut depicts a group of passengers aboard an IRT Broadway-Seventh Avenue express train. Divided fairly evenly between whites and blacks and between men and women, and even though they are riding the subway late at night, nearly all the riders are fast asleep. (MCNY)*

Mayor Fiorello H. LaGuardia posing for photographers in the motorman's cab of a BMT train, Times Square subway station, on June 1, 1940. LaGuardia was taking part in a ceremony commemorating the transfer of the Brooklyn-Manhattan transport properties from private to public ownership and the unification of the city's three separate subway systems under direct municipal control. (LaGuardia and Wagner Archives, LaGuardia Community College/CUNY)

Paul Windels. A leading Republican attorney, Windels initiated a campaign for the abolition of direct municipal subway operation, which resulted in the creation of the New York City Transit Authority in 1953. (NYHS) about $500,000 from this concession and vigorously defended its contractual rights; the company won the case several years later. The signs remained, an early indication that the subway was assuming a more pedestrian place in the life of the city.[11]

The initial thrill eventually wore off for the passengers, too, and a subway ride became nothing more than a daily habit. Elmer Rice, a twelve-year-old who had been badgering his parents to take him on the IRT, finally got his chance one Sunday afternoon late in November. Young Rice, his parents, his grandfather, and his uncle embarked on a family expedition through the subway. Years later Rice, by then an accomplished playwright, treasured this trip as a highlight of his boyhood. "So this was the subway!" he exclaimed, remembering how awed he had been by the beautiful tile mosaics and how he had pressed his face against the window glass and watched the station pillars flash by. But the excitement faded away when Rice began taking the subway to school every day. "At the end of six months," he admitted, "I had even stopped looking at the tiling."[12]

Long after most New Yorkers became bored with it, the IRT continued to fascinate out-of-towners. For years popular guidebooks put the subway high on their list of sights that travelers should see in New York. "The tourist," Rand McNally advised in 1905, "will be well repaid by a trip through the bore of this greatest of all underground railways," while the *Banner Guide & Excursion Book* strongly encouraged visitors to experience the subway's "surprising roominess and apparent rush and hustle and, especially, the fine finish and cheeriness of its stations."[13] As these guidebooks recognized, the IRT was an essential part of the metropolis. Along with the Brooklyn Bridge, the Statue of Liberty, the Flatiron Building, and Wall Street, the IRT embodied the wealth, power, and modernity that distinguished New York from all other cities.

For residents and tourists alike, the IRT represented a completely new kind of urban environment. Unlike the els and streetcars where the rest of the city was always in view, the underground railway enabled passengers to travel across New York without ever catching sight of the surface. For instance, riders could pass through the city hall subway station and never know what city hall, the Tweed courthouse, or the surrogate courthouse looked like. According to the *Utica Saturday Globe,* the subway had transformed New York into "the city of human prairie dogs." The paper noted that "just as the little burrowers of the West dart into their holes in the ground," so New Yorkers had developed a habit of disappearing beneath the surface and then reappearing somewhere else.[14]

A ride in the subway thus meant entering a separate and sometimes disorienting sphere, particularly in the long stretches between stations where the trains were shrouded in darkness. Isolated from their familiar surroundings and dependent on steel rails, track switches, electrical conduits, signals, and other mechanical devices, passengers thought of the subway as a realm of impersonal, complicated technology.[15]

Observers agreed that the IRT's most important technical attribute was its high speed. In fact, the Interborough was renowned for being the fastest urban mass transit railway in the world. The reason for this impressive performance was that it was the first rapid transit railway to have separate express and local service. Many other railways employed some kind of express service before 1904, but none could compare to the Interborough. For example, the Manhattan Railway Company's Ninth Avenue elevated route extended all the way from Cortlandt Street at the foot of the island to the Harlem River, but its expresses were confined to one part of this route, from Fourteenth Street to 116th Street, and to a single track. Consequently, Ninth Avenue expresses headed south during the morning rush hour and back uptown at night. In comparison, the Interborough's four tracks permitted permanent two-way express service, significantly improving the speed and range of its trains.[16] The Interborough expresses exceeded forty miles per hour, three times faster than the city's steam-powered els and six times faster than its streetcars.

New Yorkers heralded the Interborough in dances, advertisements, songs, and movies. Late in 1904 a dance number called the "Subway Express Two-Step" that evoked the physical movement and emotional stimulation of an IRT trip swept the city.[17] The following spring, a haberdashery located at the corner of Broadway and Chambers Street advertised a clothing sale featuring the ' "Rapid Transit' selling" of coats, shirts, and neckties at such low prices that everything would "go by 'express' " with "no 'stop' " at all.[18] A 1907 ballad, "The Subway Express," depicted the Interborough as an arena for sexual flirtation where the motion of a crowded rush hour express threw two strangers, a young man and a young woman, into each other's arms. As the train sped from lower Manhattan to the Bronx, they fell in love:

(Boy)
It was in no sheltered nook
It was by no babbling brook
When romantic'lly we met.
(Girl)
Ah, the scene I can't forget
We were thrown together in the Subway Express.
(Boy)
You were clearly all at sea
As you wildly clutched at me
When around that curve we swung.
(Girl)
Yes, and though I'd lost my tongue
I made a hit with you, you must confess.
(Boy)
Yes, you hit me in the back,
And as around and round you flew
I inquired if I could tender a supporting arm to you.
(Girl)
To which I answered "No, sir!"
When the guard yelled "Move up closer"
And clearly there was nothing else to do.

(Boy)
We first met down at Spring Street
And then upon my word.
(Girl)
I felt I'd known you all my life
When we reached Twenty Third.

(Boy)
You won my heart at Harlem
(Girl)
At the Bronx I murmured yes,
(Boy)
We lost no time in that hour sublime
On the Subway Express.[19]

The boy and the girl experienced an epiphany that changed their humdrum daily routines into something magical. As they were taking an ordinary ride home during rush hour, the sensations of speed, noise, and motion suddenly invested their trip with sexual adventure. Hurtling through the city, the train became a highly charged environment where the social customs that usually governed courtship were temporarily relaxed; men and women would touch each other physically and emotionally. In this song, if not always in reality, the technology of the subway created a stimulating but safe milieu that enriched riders' lives.[20]

But the subway's high speed could be dangerous, too. During the Interborough's early years, its fast-moving trains struck and killed an alarming number of people who were out for a walk on the tracks. Just one week after the IRT opened, a middle-aged Bronx resident with the singular name of Leidschmudel Dreispul, who was apparently captivated by the new marvel, decided to explore the tunnel for himself. Dreispul began hiking south along the tracks from the 137th Street station, but a southbound express killed him one block into his trip. Several years later an out-of-work laborer by the name of Timothy McCarthy, who was standing on the downtown platform of the Astor Place station, tried to save the price of a nickel fare by sneaking across the tracks to the uptown platform; although McCarthy managed to cross three sets of tracks, an uptown local running on the fourth set of tracks whirled him beneath its wheels. Another fatality took place in August 1910 when a train literally caught S. Silvio with his pants down; feeling the call of nature, he ducked into the Times Square tunnel to relieve himself. So many people died walking on the tracks that the Interborough Rapid Transit Company had to post warning signs: ALL PERSONS ARE FORBIDDEN TO ENTER UPON OR CROSS THE TRACKS.

These victims died because they were used to the poky trolleys and els, not the swift IRT. Many New Yorkers thought nothing of nipping across the street in front of a trolley. (According to legend, this practice of dodging street cars inspired the name of Brooklyn's National League baseball team, the Dodgers.) Even though trams killed and maimed dozens of people year after year, they moved so slowly in traffic that pedestrians could not resist the temptation. This behavior was not completely unreasonable. A man who was walking three miles an hour could cover four and a half feet per second, not much less than a trolley (which, at six miles per hour in traffic, traveled nine feet per second) or even an el (which, at twelve miles per hour, traveled eighteen feet per second). But a forty-mile-per-hour express, which traversed fifty-nine feet per second, belonged in an entirely different dimension; avoiding it was almost impossible. Anybody who failed to appreciate that the IRT tracks were an extremely dangerous corridor might pay for the error with his life.[21]

The Mighty Metropolis

In his speech on opening day, Mayor George B. McClellan boldly predicted that the subway would guarantee New York's status as a great city. Despite the unification of the five separate boroughs in 1898, McClellan claimed that greater New York remained "little more than a geographical expression" and had yet to become a cohesive economic unit. In his view the IRT would make New Yorkers the "sons of the mightiest metropolis the world has ever known" by overcoming the formidable geographical barriers that impeded interborough communication and by promoting development on the city's outskirts.[22]

McClellan's prediction came true within a decade. The Interborough's swift expresses brought vast reaches within easy traveling time of downtown, unleashing a gigantic construction boom that fundamentally altered existing neighborhoods and tripled the size of the built-up territory. So significant were the changes brought by the IRT that it virtually redefined areas as disparate as Times Square, the Upper West Side, and the Bronx.

Times Square

Times Square was one of New York's most unsavory neighborhoods before 1904. Known during the nineteenth century as Longacre Square after a similar place in London, it housed horse-related businesses such as livery stables, animal traders, carriage shops, saddlers, and harness makers. Its streets were lined with landaus, broughams, and tradesmen's carts waiting to be repaired, and its shabby buildings reeked of manure. Longacre Square had a dangerous reputation, especially after dark when prostitutes and thugs flourished. To the west lay Hell's Kitchen, a tough Irish slum whose tenements, factories, and saloons added to the square's forbidding atmosphere.

Despite its seedy appearance, Longacre Square had splendid possibilities for commercial development. Sitting astride a great transportation hub where Broadway, Seventh Avenue, and Forty-second Street met, this crossroads in the 1880s and 1890s attracted such cultural establishments as the nearby sumptuous Metropolitan Opera House and Oscar Hammerstein's Olympia Theatre.[23]

The IRT helped Longacre Square realize its potential. The first significant improvement was the Hotel Astor, a huge French Renaissance building completed at Broadway and Forty-fourth Street in 1904. Among New York's largest and most elaborate hotels, the Astor was extremely expensive, with room prices ranging from $2.50 a day for a single without a bath to $10.00 for a luxurious suite, and it soon became renowned for its fashionable balls and charity dinners. Even more noteworthy was newspaper publisher Adolph Ochs' Times Building. Early in 1902, Ochs resolved to leave the *Times'* cramped and obsolete offices on newspaper row near city hall. He bought a triangular plot on Broadway between Forty-second and Forty-third Streets where the Pabst Building stood. Ochs' move not only prompted the city to rename this district Times Square (in April 1904), but it also spawned a now-venerable custom. Always trying to get an edge over his competition, Ochs opened his Times Building —a pink granite and terra-cotta tower that was then the second tallest structure in the city—on December 31, 1904, with a magnificent fireworks display that drew tens of thousands of people. Skyrockets and flares lit up the night, and when an illuminated globe slid down the side of the tower and reached the bottom, a final burst of firecrackers signaled the beginning of the new year by spelling out *1905* above the square. After the success of this extravaganza, Times Square emerged as the site where first the city and then the whole nation welcomed the new year.[24]

The IRT also attracted theaters to Times Square. Before 1904, Manhattan's legitimate theaters had clustered around Herald Square, at Thirty-fourth Street and Broadway. Like baseball grounds, amusement parks, and vaudeville houses, playhouses required locations near the transport routes that brought customers to their doors; during the 1880s and 1890s no part of upper Broadway had better connections than Herald Square, which was served by the Sixth Avenue elevated and by several trolley lines. But the Interborough changed that. In his 1904 musical, *Little Johnny Jones,* famed songwriter George M. Cohan noted the northward migration of theaters:

> Give my regards to Broadway,
> Remember me to Herald Square.
> Tell all the gang at Forty-second Street
> That I will soon be there.[25]

This relocation was virtually complete by 1906 when the *Real Estate Record and Builders Guide* proclaimed there was "a concentration of theatrical businesses in that one area the likes of which has never been seen in New York before."[26] Times Square already boasted outstanding playhouses, such as the Astor, which produced *A Midsummer Night's Dream* for its September 1906 debut and presented the first Pulitzer Prize winner in drama eleven years later, Jesse Lynch Williams' now-forgotten *Why Marry?*

From the beginning, legitimate theaters shared Times Square with more popular entertainment forms: vaudeville houses, burlesque palaces, and music halls that catered to working-class audiences by offering livelier diversions and cheaper prices. At first the legitimate theaters occupied plots right on Times Square and overshadowed their popular rivals, but rising real estate costs and competition from the movies—a dazzling new entertainment medium that appealed to the masses—altered this pattern. Although New York's first movie houses were low-budget, storefront operations scattered in working-class neighborhoods across the city, Times Square gradually emerged as the city's prime movie quarter. Shortly before World War I, construction began on elaborate new auditoriums intended to show movies exclusively. At the same time, the first high-quality, feature-length films began to appear. In April 1914 the three-thousand-seat Strand, at Broadway and Forty-seventh Street, became one of the country's first theaters especially designed to show movies, while one year later the landmark two-and-a-half-hour film *The Birth of a Nation* had its New York premiere at the Liberty, at 234 West Forty-second Street. After World War I, fabulous motion picture palaces such as the Paramount and the Roxy went up on Broadway and on Forty-second Street, forcing the owners of legitimate theaters to either convert to the new medium or move onto side streets in the west Forties and Fifties. The triumph of the movies was confirmed in 1927 when the first satisfactory talking picture, *The Jazz Singer,* featuring vaudeville star Al Jolson, premiered at the Warner on Broadway at Fifty-first Street.[27]

By the 1920s, Times Square had evolved into a spectacular entertainment district devoted to rapid, unceasing change as a form of pleasure. Times Square brought the squalor and excitement of Coney Island's amusement parks into the heart of the city; its playhouses, movie palaces, speakeasies, restaurants, and dance halls created a zone of fantasy that was easily accessible to almost all Manhattan neighborhoods. Perhaps nothing typified Times Square's frenetic appeal more than its electric advertisements. Beginning around 1917 when the Wrigley Chewing Gum Company installed a huge, four-story-high electric sign on the west side of Broadway from Forty-fourth to Forty-fifth streets that showcased a gurgling fountain and a pair of colorful peacocks, advertisers filled the square with spectacular displays that lit up the night and magnified the square's hallucinatory atmosphere. A second Wrigley masterpiece portrayed a pack of spearmint gum floating in an ocean of gurgling fish, and a risqué Bond Clothing billboard showed a semi-nude man and woman separated by a coursing waterfall.[28]

The IRT subway was largely responsible for this new Times

Square. The square's movie houses and theaters depended on the subway to carry huge volumes of passengers to their ticket booths. Perhaps equally important were the IRT's hours of operation. Open twenty-four hours a day soon after its inauguration, the IRT helped give Times Square—and New York City itself—a reputation as a place that never closed, a city where anything was always possible.[29]

The Upper West Side

Born on October 19, 1895, Walter Mack spent his childhood at 312 West Seventy-first Street on a prosperous upper-middle-class block between West End Avenue and Riverside Drive. At the time the neighborhood was a suburban haven far removed from the noise and confusion of downtown Manhattan. Writing his autobiography years later after he retired as president of Pepsi-Cola, Mack recalled how quiet his street had been at the turn of the century: On some summer days the only disturbance was a horse-drawn cart clattering while delivering milk, ice, bread, or coal from door to door. With so little traffic to worry about, parents let their children play stickball and tag in the middle of the street. In Mack's eyes the Upper West Side was hardly part of the city at all. "Each block of brownstones was more like a small village unto itself than part of a thriving, prosperous, driving city going through growing pains," Mack wrote, "and everyone had the time for neighborly pleasantries."[30]

What Walter Mack recalled about his block was true of the approximately two hundred other blocks that comprised the Upper West Side. As late as 1900 this sprawling district—which went from Fifty-ninth Street to 110th Street and from Central Park West to the Hudson River—was not fully developed. Because the Ninth Avenue elevated was the only rapid transit railway, most of the Upper West Side's buildings were located along its Columbus Avenue route (above Fifty-ninth Street, Ninth Avenue became Columbus Avenue) or on the two nearest boulevards, Central Park West and Amsterdam Avenue. When urban boosters such as C. T. Hill, who promoted the area in an 1896 *Harper's Weekly* article, pointed to "the rapid and notable improvements" going up on the Upper West Side or to "the armies of people'' who were swarming there, it was this stretch from Amsterdam to Central Park West that they had in mind.[31] The near west side had grown rapidly during the 1880s and 1890s; builders such as Edward S. Clark, Bernard S. Levy, and Edward Kilpatrick erected upper-middle-class row houses on the side streets below Seventy-ninth Street and elegant apartment houses such as the Endicott, Nebraska, and Lyndhurst on Columbus Avenue. In addition to this housing for the affluent, many tenements arose on Amsterdam Avenue and on the side streets above Seventy-ninth Street.

By contrast the western avenues—Broadway, West End, and Riverside—were largely neglected due to poor transport. Loath to walk ten or fifteen minutes to the nearest Ninth Avenue el station or to endure the slow, unreliable Broadway streetcars, homebuyers generally avoided this section. To be sure a number of private homes did go up on the far west side after 1885. For instance, in 1893–94 the speculative contractors J. and D. Dunn erected eleven one-family brownstones in the three hundred block of West Eighty-fourth Street. Sold for about $16,000 apiece, these homes provided upper-middle-class families with comfortable living arrangements that rivaled anything in downtown Manhattan. But modern dwellings like the Dunns' row houses were the exception rather than the rule. As late as 1900 over half of the far west side lots were a collection of empty rock-strewn earth, farm gardens, and shanties, together with the ruins of a few dilapidated colonial-era mansions.[32]

Broadway was in particularly bad shape. For decades observers predicted that this elm-lined boulevard would become a grand residential thoroughfare like the Champs Elysées, with finer mansions than Fifth Avenue's. But precisely because of Broadway's exceptional promise, speculators gained control of most of its frontage. These speculators were attracted to the possibility of windfall profits, and they refused to sell or improve their property until land values doubled or tripled. As a result Broadway stagnated. More than one-third of the street corners from Seventy-sixth to Ninety-sixth streets were empty in 1898; others were occupied by ugly coal yards and lumberyards.[33] ''Broadway presented anything but an attractive view,'' the *Real Estate Record and Builders Guide* recalled. ''The vacant property was in most instances surrounded with broad fences, usually covered with unsightly posters, and the small buildings were anything but architectural monuments."[34]

Urban philosopher Lewis Mumford, who grew up on West Sixty-fifth Street around the turn of the century, vividly recalled the boulevard's raw and tattered appearance. "From Sixty-fifth Street up," Mumford wrote in his autobiography, "Broadway was still full of vacant lots, with visible chickens and market gardens, genuine beer gardens like Unter den Linden, and even more rural areas."[35] Young Mumford often took long walks through the Upper West Side with his grandfather, and its open spaces and leafy promenades made a deep impression on him. In particular Mumford remembered an empty plot on the old Astor estate in the lower Nineties where squatters lived in tumbledown shacks and tended thriving market gardens. Indeed, Mumford claimed that his happy boyhood memories influenced his design preference for cities such as Rome, Paris, and London that had low population densities and broad vistas.[36]

With the arrival of the subway, the Upper West Side began to change. A construction boom began that raised the assessed value of taxable land west of Amsterdam Avenue 33 percent from 1905 to 1913, compared to only 11 percent east of Amsterdam.[37] The district's three main thoroughfares acquired distinctive personalities. A phalanx of huge, multistory apartment buildings rose on West End above Seventy-second Street, turning it into a solidly middle-class avenue with shade trees, empty sidewalks, and blank windows that communicated an air of bourgeois respectability. More impressive edifices went up on Riverside Drive. A winding, dipping road that hugs Manhattan's western ridge line for almost seven miles from Seventy-second Street to Dyckman Street, Riverside Drive commands spectacular views of the Hudson River and the Palisades. Over the years bicyclists and carriage drivers made this heavily wooded promenade a favorite leisure ground, and splendid public memorials, such as the Soldiers' and Sailors' Monument to the Union War dead at Eighty-ninth Street, were put there. Sloping gently from Riverside Drive to the Hudson was Frederick Law Olmsted's Riverside Park, where trees, shrubs, and flowers were planted in the pattern of an English garden. As a result of these scenic amenities, unusually fine apartment buildings went up on Riverside Drive, offering luxurious suites of five to ten rooms and plenty of space for servants. Beyond question the ritziest flat was newspaper publisher William Randolph Hearst's penthouse, which occupied the three top floors of the Clarendon at Eighty-sixth Street and Riverside Drive. It included over thirty rooms and totaled more than three-quarters of an acre of living space.[38]

But no street changed more than Broadway. With the nine IRT stations located between Fifty-ninth and 110th streets discharging tens of thousands of passengers onto its sidewalks every single day, Broadway quickly replaced Columbus Avenue as the Upper West Side's chief artery.[39] After 1904 big ten- to fourteen-story elevator-equipped apartments sprouted on Broadway, their ground floors housing retail businesses that catered to the heavy pedestrian traffic, their upper floors reserved for apartments. The Belnord at Broadway and Eighty-sixth Street was fairly typical. Built around a spacious, well-landscaped courtyard, the thirteen-story Belnord occupied an entire city block and ranked as the largest apartment building in the world when it was completed in 1909. Its 175 flats, ranging in size from seven to fourteen rooms, were ornamented with elaborate touches like mahogany doors and wainscoting. At a time when working-class families paid from $7.00 to $20.00 monthly for their spartan accommodations, rents at the posh Belnord began at an incredible $175.[40]

The IRT made the Upper West Side a densely populated, heavily built-up urban center. According to one longtime resident who missed the more relaxed, personal life of the 1890s, by 1921 the traffic on Broadway was so thick that police officers had to be stationed on the corners of Seventy-second and Eighty-sixth streets to help pedestrians cross the street. For better or for worse, the Upper West Side was now part of the metropolis.[41]

The Bronx

A photograph taken in 1910 captured a critical moment in the Bronx's history. The picture reveals open fields in the Soundview section of the eastern Bronx where the streets department had laid out ribbons outlining the roads and avenues it planned to build. Just a few years earlier this part of the borough was farmland where crops such as potatoes and cabbages grew; by 1913 it was ready for urban development: The streets were opened, the property subdivided into fifty- by one-hundred-foot building lots, and water and sewer pipes laid. Because this area was only a trolley ride away from the IRT's East 180th Street stop, its development seemed assured. All that remained was for building contractors to erect houses and stores there.[42]

Today farms, villages, and country estates; tomorrow, a city. From 1904 to World War I this process of rapid, comprehensive urban development was repeated throughout the Bronx in neighborhood after neighborhood along the IRT's Broadway and Lenox Avenue branches. Before the IRT arrived, the territory west of the Bronx River was more heavily built up than the eastern Bronx; in 1900 it held most of the borough's two hundred thousand residents. The subway set off a wave of construction that engulfed both sections.[43] One Bronx native who witnessed the IRT's dramatic impact on the rural landscape was Bert Sack, a longtime resident of Hunt's Point in the southwestern Bronx. Born in 1896, Sack grew up in the countryside. In an oral history interview sixty years later, Sack remembered picking pears, cherries, and peaches in orchards near his home and having to walk ten blocks to find a place that made ice cream sodas. He also remembered how his village's rural atmosphere came to a sudden end:

> In 1903 I watched the building of the elevated structure of the subway on Southern Boulevard from our window on Fox Street. The next year, 1904, the subway started operating, but not in the Bronx. The terminus was at 145th Street and Lenox Avenue. In the meantime, the Second Avenue el came by a spur at 149th Street as far as Freeman Street. Finally, about 1905, through service was opened to Bronx Park by subway. This was the beginning of the end of the Bronx as the home of farms and estates. The Tiffanys, Foxes, and all the other old families soon abandoned their homes. Many streets, such as Fox, Tiffany, and Kelly, which had ended at Westchester Avenue, were cut through to Intervale Avenue. A Sunday pastime then was the inspection of the new flats being built. . . . It did not take long before the rest of Simpson, Fox, Tiffany, and Kelly streets were filled with flats. The change had started.[44]

Bigtime real estate speculators were the agents of this change. They gained advance knowledge of the subway's route from Interborough executives and municipal politicians, bought vast tracts of unimproved land along the line, and then sold their property to building contractors at tremendous profit. Although speculators operated everywhere along the IRT, they were particularly important in northern Manhattan and the Bronx. Most of the property in lower Manhattan had already been broken into small lots, but above 125th Street speculators could make a killing by buying a single farm or golf course and then waiting for prices to climb. One or two speculators could thus have a powerful impact on an entire district.[45]

New York's top speculator was Charles T. Barney. The son of a Cleveland businessman, Barney graduated from Williams College in 1870, moved to New York City, and in 1875 married the sister of financier William C. Whitney. With this marriage clearing his way to enter Wall Street's highest circles, Barney became president of the Knickerbocker Trust Company, a special partner in the brokerage firm of Rogers and Gould, and the director of several real estate and insurance firms. An active land speculator, in 1891 he formed an investment combine that bought property along a planned subway route through Washington Heights in upper Manhattan.

That scheme collapsed, but Barney got a second chance when he became a director of the IRT subway in 1900. To exploit his inside knowledge about the subway's route, station locations, and construction timetables, Barney organized a syndicate that commanded nearly $7 million in capital. This syndicate began buying property in Harlem, Washington Heights, and Fort George around 1901, and it shifted to Inwood and the Bronx the following year. For instance, it purchased 109 lots around the 225th Street and 238th Street stations in the Kingsbridge section of the Bronx. No inventory of the syndicate's holdings was ever made public, but Barney held title to 350 parcels in his name alone and probably made hundreds of thousands of dollars.[46]

The builders who bought land from Barney encountered severe financial pressures to develop and sell their properties rapidly in order to recoup their investments and pay back their loans. Consequently, these builders turned to a type of housing that could be erected quickly, cheaply, and for a ready market, while still offering good income relative to the high land costs.[47] The most common form of residential building in the Bronx was the new-law tenement. From 1904 to 1914, 6,373 new-law tenements were erected in the Bronx, most of them along the IRT's Broadway and Lenox Avenue branches. This concentration of tenements formed—in the words of the *New York World*—"a distinctive subway zone of flat houses" that made the Bronx synonymous with low-income housing.[48]

But the quality of this housing was relatively high. A substantial structure that rose six stories and contained about thirty apartments, the new-law tenement was named for a 1901 municipal reform that banned the old-law tenement, which was notorious for depriving tenants of light and air and for fostering poor sanitary conditions. Filthy, ramshackle old-law tenements blanketed the Lower East Side, Greenwich Village, and Hell's Kitchen. By contrast the new-law tenements going up in the Bronx seemed wonderful. For $16.00 to $20.00 dollars a month residents could rent a brand-new apartment that included two bedrooms, a combined dining-living room, a kitchen with hot water and a gas range, and an interior toilet and bathtub. These new-law tenements had good heating and lighting, carpeted hallways, and tastefully decorated foyers and facades.[49]

To working-class New Yorkers who longed to escape from Manhattan's crowded slums, the Bronx represented a way station on the road to the middle class. Although the poorest could afford neither the time for commuting nor the rents, the Bronx was a possibility for most semiskilled and skilled workers. Thousands of people, particularly German Protestants and German and Russian Jews, migrated to the Bronx after 1904 and settled in solidly ethnic communities such as East Tremont and West Farms. In 1920, 75 percent of the residents who lived in the census tracts that embraced the original IRT subway had either been born abroad or were the children of foreign-born parents. For this generation of Jews, Germans, Irish, and others who combined strong group identity with individual striving for the better life, the Bronx was a promised land.[50]

The IRT changed New York City almost beyond recognition. Gone were the farms and villages that had covered northern Manhattan and the Bronx in 1900, and gone was the pattern of sparse and uneven development that had characterized the Upper West Side. From Times Square's theaters and movie houses to the Upper West Side's huge ten-story apartments and to the new working-class neighborhoods in the Bronx, the face of the city acquired a new set of distinctly urban features.

The settlement of the outlying sections of northern Manhattan and the Bronx was the most important consequence of the IRT. In effect the subway represented an indirect municipal subsidy to the private construction industry that built this new housing and to the low-income tenants who inhabited it. Only a rapid transit railway as fast as the IRT could have opened up these distant regions for residential expansion, and only a publicly financed subway could have achieved the IRT's high technological standards. Without violating the laissez-faire taboo against direct intervention in the private sector, city government helped provide decent accommodations for working-class families. The subway's enduring legacy was that the lives of New York's poorer citizens became fuller and more productive.[51]

5

GOOD-BYE TO THE PATRICIANS

The Subway Crush

The day after the IRT opened in October 1904, the *New York Tribune* announced "the birth of [the] subway crush." So many people wanted to ride the new railway that the cars, platforms, ticket lines, and kiosks remained mobbed no matter how many trains the Interborough Rapid Transit Company rushed into service. After seven more days of unusually heavy traffic, the *Real Estate Record and Builders Guide* concluded that "the subway should have been designed to handle much larger crowds than existing stations and their approaches can handle."[1]

The *Record and Guide* was right. The principal reason for this congestion was that the IRT's expresses proved to be far more popular than anyone had anticipated. Contrary to the engineers' forecasts, New Yorkers were in such a hurry to get where they were going that the expresses were far more crowded than the locals. These fast-moving trains also boosted subway patronage by triggering residential settlement in outlying areas. From 1905 to 1920 the population of Manhattan above 125th Street grew 265 percent, to 323,800, and the population of the Bronx advanced 150 percent, to 430,980. These new neighborhoods funneled more and more passengers into a subway that was already overflowing.[2] "At the present time," the *North Side News* declared as early as February 1906, "the trains and cars of the subway and elevated are jammed night and morning with people who want to go to their homes."[3]

Instead of simply diverting passengers from Manhattan's elevated and surface railways, the subway stimulated ridership by encouraging New Yorkers to travel more frequently. Between 1904 and 1914 the average number of rides that each New Yorker took on public transport per year jumped from 274 to 343. During the same period New York's total mass transit patronage rose more than 60 percent, to 1.75 billion. The result was that traffic steadily worsened, particularly on the IRT. The subway was designed to carry no more than six hundred thousand riders per day, but its average daily patronage surpassed that number by October 1905, only a year after opening. By 1908 the underground averaged eight hundred thousand people per day, one-third above its planned maximum capacity.

Overcrowding was particularly severe during rush hour. A traffic survey taken in 1908 showed that one-third of the IRT's daily ridership was concentrated in the two-hour peak periods of 8:00 A.M. to 10:00 A.M. and 4:00 P.M. to 6:00 P.M. Obviously, not everyone could find a seat then. More than half of the twenty-eight thousand southbound riders who passed through the IRT's Grand Central station one morning in November 1908, for example, had to stand and hang on to the straps.[4]

The Interborough Rapid Transit Company tried to keep pace with this surge by modifying the subway to hold more people. It extended station platforms so that longer trains could be run, designed a new kind of car that had three doors on each side instead of two so that passengers could exit and enter more quickly, and adjusted the signals to shorten the headway between trains and move more trains through the route. But although the IRT managed to squeeze more riders below ground, these increases took place without a corresponding expansion of trackage, and the trains remained swamped. The subway was probably more crowded before World War I than it has been since then. Indeed, the IRT's traffic was denser than that of any rapid transit line in the world. In 1905 the IRT bore 3.6 million people per mile of track, twice as many as the

New York elevateds did. In 1914 the IRT carried 9.5 million people per mile of route, far more than the Paris metro (7.2 million), the Berlin railway (5.6 million), and the London underground (4.4 million).[5]

Yet passenger statistics did not tell the whole story. In addition to the sheer volume of riders, the renowned individualism of New Yorkers intensified the traffic crunch. The behavior of transit passengers varies markedly from culture to culture; subway crowds in New York act differently from those in London, Tokyo, Mexico City, Vienna, and Moscow. For example, Londoners patiently queue at ticket booths and bus stops; divide into two columns on the escalators, those who want to stand in place forming one line on the right-hand side of the escalator, and those who prefer to keep walking forming another line on the left-hand side; and move through clogged underground corridors almost without touching one another. Unlike the English, who were confined to a small island, Americans enjoyed plentiful open land and low population densities on a spacious continent and did not need to subordinate their individual identities for the sake of social harmony. Partly for such cultural reasons, New Yorkers in their subways behaved much differently from Londoners in their tubes. Even in 1904, New Yorkers apparently felt little social pressure to refrain from pushing, shoving, seat-stealing, or line-cutting. Ironically, such disorderliness actually aggravated the traffic problem by impairing subway efficiency: Passengers who delayed trains at stations beyond schedule reduced the IRT's carrying capacity.[6]

Hardly anyone enjoyed rush hour. To use the subway, patrons walked down steep, narrow stairwells, waited in long lines to buy tickets, and then entered onto dirty, poorly ventilated platforms.[7] The poor condition of the stations was due partly to several filthy masculine habits. Rider Erwin de Kohler complained: "I have become accustomed to the crowding and the pushing and all the discomforts that are the natural consequence of the herding of hundreds of people in a confined place, but there is one thing I cannot tolerate . . . [and] that is . . . 'spitting.' "[8] Despite the municipality's efforts to stop the spread of tuberculosis by banning the traditional American male habit of spitting in public, some men refused to change their ways, and the floors were covered with gobs of spit. Similarly, even though the Board of Health had prohibited smoking in the subways, a passenger named Whidden Graham griped that all too many men showed a "hoggish disregard for others' rights" by puffing away on their "vile cigarettes, cheap cigars, and evil-smelling pipes."[9] "The stench emanating from some of the pipes and dead nickel-a-box cigars . . . has been enough to make me wish that the Tobacco Trust had been legislated out of existence," another male passenger grumbled. "In the Bronx subway I have sat next to pipes [that] . . . give forth such a rancid, noisome odor as to make one think an asafoetida bomb had been exploded someplace on the train."[10]

After the train arrived at the station, passengers tried to enter the cars in a mad scramble that reminded some people of a fullback's dive into the line of scrimmage. Hildegarde Hawthorne, an English travel writer, described the scene at one busy IRT platform in 1911 as

> a sight not to be matched elsewhere—happily! As each train pulls up and the doors are slid back, a frantic rush is made by the waiting crowds. "Let 'em off!" yells the guard; "Keep back!" roar the police; and those who must get out push and struggle against the advance throng of those who want to get in. The policemen pull and shove; the gongs sound incessantly; and as the last passenger squeezes off, a resistless mass of humanity is wedged into each car; the doors are slid shut again, at the imminent risk of crushing someone; the policemen haul back those who have not managed to board the train, and off she whirls, to be followed the next moment by another, where the same scene is repeated.[11]

According to one disgruntled onlooker, the logical outcome of this wild struggle was for a fistfight to break out at the entrance to every car—with the winners getting the opportunity to board.[12]

Conditions were not much better inside the dark red cars. Although the Interborough had already introduced an innovative all-steel design that became the industry standard, its first fleet of five hundred cars were more traditional composites that featured steel carriages and frames and wooden floors, sides, and facings. These composites had fifty-two rattan-covered seats that were arrayed longitudinally near the ends and crosswise in the center. Dimly lit by twenty-six incandescent bulbs, the interiors were dark and bleak. The cars were very uncomfortable during peak periods when riders competed for seats or hung on to the leather straps, trying to keep their footing on the maple-slatted floors. The straphangers were crammed rudely together with embarrassingly little room to spare, shoulder pressing shoulder, hips jamming buttocks and crotches. The discomfort of the straphangers was aggravated because many passengers refused to move to the center after entering a car; instead they remained near the doors at the ends of the car in order to be the first to exit. This antisocial behavior appears to have been especially common on the expresses. The faster the trains moved, the more quickly New Yorkers wanted to get to their destinations and the more intolerant they became of even the slightest delay. The result was that thick clots of passengers formed around the doors, compelling other riders to muscle their way in and out. No wonder cynics compared the trains to sardine cans and cattle cars.[13]

Subway cars were notable for their social diversity. Because the IRT route passed through such a broad spectrum of communities, Interborough trains collected people of all ages, races, nationalities, and classes, and both sexes.[14] Years later author Kate Simon described the transit vehicles as New York's "only melting pots" because they were among the few places where every kind of urban resident mingled on a regular basis. Noticing this same phenomenon, a few commentators voiced the hope that New Yorkers would acquire greater cultural understanding just by riding the subways. Poets Langston Hughes and Joyce Kilmer, for instance, believed that the city's racial and class divisions, might be bridged during rush hour when people who ordinarily avoided each other came face-to-face. Thrust together so intimately that they had to see one other, perhaps for the first time, strangers would finally have to learn to overcome their fears and recognize their common humanity. In reality, however, the subway probably exacerbated social tensions. Overcrowding was a perfect recipe for anxiety. Many riders resented the personal indignity of the subway crush, and others disliked being trapped below ground in a small, confined space where they were unable to see the daylight or readily escape to the surface. The subway could also be a pressure cooker—the ethnic, gender, and class differences sometimes led to explosions.[15]

Perhaps nothing vexed contemporaries more than the sexual harassment of women and girls. In 1912, *Outlook* magazine complained that female riders endured "a crowding which at best is almost intolerable and at its worst is deliberately insulting." Exploiting the lack of space during peak periods, males "who were often not too chivalrous, and sometimes coarse-grained, vulgar or licentious," fondled women almost at will. Women accounted for only about a quarter of all rush hour passengers, and many of them seemed to have been "jostled" or "insulted," as unwelcome sexual contact was called. Although sexual abuse as a general phenomenon was not confined to a single class or ethnic group, *Outlook* was primarily worried about one particular pattern: the molestation by working-class men of young middle-class girls. To *Outlook* such cross-class sexual contact amounted to "a violation of the laws of decency" constituting the most serious condemnation of subway overcrowding.[16]

Others were equally outraged. Three years earlier, in February 1909, Julia D. Longfellow, the wife of a distinguished attorney and a leader of the Women's Municipal League, proposed that the last car of every IRT train be reserved exclusively for women during rush hour. Although Longfellow thought that women should be able to ride anywhere in the trains, as in the past, she claimed that a separate, male-free space should be set aside for those women traveling alone and unprotected who were "physically unable to cope with the fearful crushes" of peak periods, who had to ride with young children or carry bundles, or who had suffered from sexual "insults and indignities which they have been powerless to avoid."[17] The Interborough flatly rejected Longfellow's idea, but the Hudson and Manhattan Railroad (H&M), which operated a subway that ran under the Hudson River from New York City to New Jersey, adopted it. Beginning on April 1, 1909, the H&M reserved the last car of its trains for women between 7:30 and 9:00 in the morning and between 4:30 and 7:00 in the evening. The ladies' cars transported large numbers during the first few weeks, especially in the afternoon when shoppers were only too happy to let the Hudson and Manhattan's special redcapped attendants carry their packages. Yet the new service was strongly opposed by other women who feared that such so-called special privileges might erode their existing limited rights. One woman scoffed: "Get into a suffragette car? Never! I am no better than the men, and the cars that are good enough for them are good enough for me."[18] Another rider, who belonged to the Equality League of Self-Supporting Women, belittled the patrons of the ladies' car as a collection of "timid females," "man haters," and "shrinking creatures," and contended that independent women could hold their own in the rush hour stampede.[19] For several months the critics of the ladies' cars engaged in a running dispute with Julia D. Longfellow and her allies. Then the whole issue became moot: The Hudson and Manhattan discontinued the ladies' cars on July 1, 1909. Although the cars had been well patronized at the outset, the H&M reported that passenger numbers had steadily diminished and no longer justified the special service. From then on women had to fend for themselves on the H&M, as they always did on the IRT.[20]

Few aspects of rush hour congestion were more annoying than sexual abuse. Yet many New Yorkers who did not experience such mistreatment also became disgusted with the miseries of subway commuting. In addition, businessmen complained that their employees were not arriving at work on time. This widespread dissatisfaction acquired a political focus when New York City's highly competitive newspapers picked up the issue and began demanding that overcrowding be alleviated. To the *Tribune,* the *World,* the *Times,* and other dailies, the solution to the subway crunch was clear: More subways had to be built.[21]

New Subways

The Rapid Transit Commission was eager to comply. Initially established to promote railway construction, the RTC continued to be expansionist minded over the years. Following the consolidation of greater New York in 1898, the RTC planned the first subway outside of Manhattan and the Bronx, the Contract No. 2 tunnel to Brooklyn.

The IRT's success reinforced the RTC's expansionism. Heavy

traffic dispelled earlier fears that underground rapid transit would be unpopular and suggested that future lines could be built on a solid financial footing. Writing in *Iron Age,* transit expert S. D. V. Burr said that

> the subway was adopted so quickly by the people and has been so successful during the first year of operation that it is quite probable that there will be some competition when bids are asked for extensions to the system. Instead of the city having something to sell that no one will buy, there will be demand for privileges to be granted.[22]

Apparently the days of having to hunt for investors were over.

The RTC wasted no time. In March 1905 a committee of the Rapid Transit Commission released a gigantic $250 million subway plan that dwarfed the original IRT. The new proposal contained nineteen separate lines that covered every borough except Staten Island and totaled 165 route miles, compared to only 22 route miles for the 1904 subway. There were to be four new trunk lines from the Battery to either upper Manhattan or the Bronx; cross-town subways along Fourteenth, Twenty-third, Thirty-fourth, and Fifty-ninth streets in Manhattan; and three new East River crossings to Brooklyn and Queens.[23]

The RTC probably did not intend to complete all nineteen routes, however. It adopted some of the lines in response to pressure from groups of local businessmen who had watched the IRT's tremendous impact on land development and who were now lobbying for subways for their own neighborhoods. Demands for rapid transit came from real estate promoters throughout the city, including the Washington Heights Taxpayers Association and the West Side Rapid Transit Association in Manhattan; the Bensonhurst and Bath Beach Subway Association and the Prospect Hill Citizen's League in Brooklyn; and the North Side Board of Trade and the Bronx League in the Bronx. Because these realtors had so much political muscle, the RTC sought to avoid a showdown by laying out almost every feasible line whether or not it really stood a chance of being built.[24]

The Rapid Transit Commission also designed some routes to protect itself from the Interborough Rapid Transit Company. Although Alexander E. Orr, Charles Stewart Smith, John H. Starin, and the other commissioners still counted on working closely with the Interborough in expanding the system, they were concerned that the company might use its control of New York's only existing subway as leverage during the negotiations for new lines. Accordingly, the RTC drafted two different types of subway routes. The first type consisted of links to the existing subway and could be operated only by the Interborough; thus, there was to be one extension to the east side IRT from the Grand Central station up Lexington Avenue to 127th Street and another extension to the west side IRT from Times Square down Seventh Avenue to the Battery. The second type of route would not be connected with the Interborough's network and could therefore be run by another transit corporation; for instance, the subways going up First, Third, Eighth, and Ninth avenues would be completely independent of the IRT. If Belmont tried to block construction or exact outrageous financial concessions, these independent routes would give the commissioners an insurance policy.[25]

The RTC encouraged regional transit companies to bid for its lines. Neither the Hudson & Manhattan Railroad nor the Brooklyn Rapid Transit took the bait, but the RTC did manage to enlist another major company, the Metropolitan Street Railway. The RTC was talking with both the Metropolitan and the Interborough by March 1906. Some commissioners concluded that this unfolding corporate rivalry would guarantee prompt subway construction. City Comptroller Edward M. Grout, who was an *ex-officio* Rapid Transit Commissioner, confidently predicted that "the companies may fight it out among themselves."[26]

But August Belmont had his own ideas.

The Traction King

August Belmont took great personal pride in the subway. Belmont rode the IRT regularly and noticed every detail about its operation. If a motorman brought his train to a rough stop when entering the station or if maintenance men turned his car into a steam bath by forgetting to open its ventilators, Belmont made sure that someone heard about it right away. He also wrote personal replies to riders who complained about rush hour schedules or rude employees. No doubt Belmont discomfited his subordinates, but he kept them on their toes and made them accountable for foul-ups.

Even more important was Belmont's recruitment of talented people to manage the subway. Most of his lieutenants were veteran railroaders who had thoroughly mastered transit operations and knew how to lead men. The Interborough's first general manager was Frank M. Hedley, the offspring of a skilled English working-class family who counted the inventor of a pioneering 1813 steam engine, the Puffing Billy, among his forebears. Hedley emigrated to the United States in 1882 at age eighteen and spent the next two decades working on elevated railways in New York and Chicago. As general manager of the Interborough beginning in 1904 and as its president from 1919 to 1934, Hedley introduced a number of mechanical improvements that upgraded the subway, including automatic speed controls, automatic doors, anti-telescoping devices, and better electric brakes.

Before becoming IRT superintendent in 1904, Abraham Lincoln Merritt (who was born on the same day in November 1860 that the nation's sixteenth president won election to office) worked for the Manhattan Elevated Railway for two decades, first as a telegraph operator and then as a dispatcher. A dignified, almost unapproachable taskmaster who closely resembled a navy petty officer or an army sergeant, Merritt was responsible for daily subway operations. He knew that "trivial" details often spelled the difference between success and failure. It was Merritt who double-checked that the station stairwells were unobstructed, that the elevators were running smoothly (just in case traffic jammed or a fire broke out), that the snowplows were ready for sudden winter storms, and that enough trains were scheduled for the annual New Year's Eve celebrations in Times Square or big baseball games at Hilltop Park (where the Highlanders—the forerunners of the Yankees—played). Merritt served as superintendent of the Interborough for thirty-six years, retiring in 1940 at seventy-nine.[27]

Thanks to Belmont, Hedley, and Merritt, New York's subway ranked among North America's best-run transit systems. It attained a good passenger safety record and a high level of mechanical reliability. Despite this solid performance, however, August Belmont opposed the construction of new subways aimed at alleviating overcrowding. On November 8, 1904, only twelve days after the inauguration of the IRT and the birth of the subway crush, Belmont bluntly told RTC president Alexander E. Orr that another building program was out of the question:

> Crowding during rush hour is inevitable. If a day ever comes when transportation during rush hours is done without crowding, the Companies doing it will fail financially. . . . There is a very fine line between success and failure. The City is always likely to grow up to its facilities in the end, if they are not established too far in advance of their needs. When in advance of their needs,—failure is the fate of the original investor.[28]

August Belmont opposed expansion because his Interborough profited from overcrowding. Heavy traffic supplied more income for roughly the same capital and operating outlays; according to an old transit maxim, "The profits are in the straps."

The Interborough's profits were huge. From 1904 until the end of World War I, the IRT was a gold mine. As early as January 1905 the Interborough Rapid Transit Company made an 8 percent profit. These earnings were truly "remarkable"—as Belmont bragged to the Rothschilds—because the company's treasury was still being drained by the costs of starting up the Contract No. 1 route and digging the Contract No. 2 tunnel.[29] That was only the beginning. Interborough earnings rose close to 20 percent at times during the prewar period, and its dividends soared from 2 percent in 1904, to 9 percent in 1907, and to 15 percent in 1915.[30]

Although August Belmont was reluctant to build almost any new subway, he particularly objected to the Rapid Transit Commission's 1905 blueprint. Instead of restricting its new lines to built-up areas that could generate big passenger loads, the 1905 RTC plan contained a number of far-flung, lightly traveled routes that were certain to dilute the Interborough's revenue and add to its debt. Moreover, Belmont was unhappy that the RTC invited competing transit companies such as the Metropolitan to bid for the new routes. He regarded the New York subway as his domain alone.[31]

Although Belmont began talking with the Rapid Transit Commission in March 1905, he was probably bargaining in bad faith and never meant to come to terms. Three years earlier Belmont had responded to a similar threat from the Manhattan Railway by acquiring that company and reorganizing its four elevateds as a division of the Interborough. He was not going to let anyone push him into unwanted expansion now, either. So a few days before Christmas 1905 the Interborough Rapid Transit Company took over the Metropolitan Street Railway and created a powerful new $220 million holding company known as the Interborough-Metropolitan.[32] The Interborough thus eliminated its only rival for control of the Manhattan railways and became the unchallenged master of the island's street and elevated railways and New York's only subway.

News of the merger stunned the city. BELMONT IS TRACTION KING; BELMONT NOW IN POSITION TO SANDBAG CITY the *New York Tribune* screamed. NEW SUBWAYS ALL AT THE MERCY OF BELMONT the *New York American* warned.[33] Coming at the crest of a huge merger wave that transformed the U.S. economy between 1897 and 1907 by putting industry after industry in the hands of corporate giants such as U. S. Steel and American Tobacco, the subway consolidation aroused widespread anxiety that big business would swallow local transit. In the phrase of the *New York Evening Post,* the Interborough represented "a new engine of political power . . . an intrenched and insolent monopoly" that threatened to overrun New York City.[34]

Orr, Starin, Smith, and the other commissioners did not know what hit them. Just a few days earlier the commissioners were confident that their divide-and-conquer strategy would produce new subways from one company or the other before too long. Now they confronted a single, all-powerful monopoly that was in no mood to dicker about anything.

The commissioners lacked the statutory powers required to deal with this changed situation. They were handcuffed by the Rapid Transit Act of 1894, which had incorporated the prevailing laissez-faire thinking about minimal government. Because this act emphasized rapid transit development rather than operation, the RTC could not regulate the subway adequately, and it had no jurisdiction whatsoever over the Interborough's elevated and surface divisions, which the state railroad commission supervised. Because the act prohibited municipal operation, the RTC could not use that possibility as a means of forcing the IRT to the negotiating table.[35]

Even if the RTC had been able to acquire more power, however, the commissioners probably would not have used it. Orr, Starin, and Smith would have considered a formal regulatory posture, as we know it, alien and almost unthinkable. As products of the nineteenth-century mercantile domination of city affairs, the commissioners preferred to make decisions through private conversations with fellow members of the commercial elite. They liked to reach consensus in an informal, clubby atmosphere where everyone knew one another and there was no room for professional politicians, scientific experts, newspaper reporters, and the general public. Yet the merchants' personal style was clearly out of step with the growing complexity of twentieth-century New York. Although small, highly cohesive business elites continued to exert great influence decades later in cities such as Pittsburgh (through the Duquesne Club), Atlanta (the Capital City Club), New Orleans (the Boston Club), Dallas (the Citizens' Council), and San Francisco (the Bohemian Club), New York City's commercial aristocracy was losing its grip by 1900. This change in power structure resulted largely from the overwhelming size of greater New York as well as from the dynamism of a complicated metropolitan economy that produced multiple, competing business elites in sectors such as communications, retailing, manufacturing, commerce, and banking.[36]

There were specific reasons why the commissioners were having such a hard time now. Although the merchants had succeeded in tightly controlling the planning and construction of the subway, the Interborough's opening changed the political landscape. Now that hundreds of thousands rode the subway every day, now that average New Yorkers had a stake in the IRT, decisions could no longer be confined to a tiny group of blue bloods. As the angry press reaction to the Interborough merger revealed, subway politics now involved the mobilization of the general public and the highly visible participation of newspapers and elected officials.

Subway politics had changed so much that the rapid transit commissioners no longer seemed very formidable. Increasingly, the commissioners' enemies ridiculed them for being old-fashioned and ineffectual. August Belmont sneered that the Board of Rapid Transit Railroad Commissioners was "composed of very respectable old gentlemen who are too timid to take a decided stand." Curiously, one of Belmont's opponents, muckraker Ray Stannard Baker, employed virtually the same language to attack the RTC. In a 1905 *McClure's* magazine article, Baker belittled the commissioners as "fine, high-minded, soft-hearted old gentlemen" who represented "all that was respectable, stable, old" and who conveyed a "halting, timid, compromising air." According to Baker, these "fine old gentlemen" were simply not virile enough to protect the city from the Interborough's rapacious executives, those "alert, grim-jawed young men who will stop at nothing—nothing—in getting what they want, and the public be damned!"[37] Belmont and Baker had a point: These "gentlemen" were out of their element in the new, more intensely politicized environment.

Suddenly the Rapid Transit Commission was in deep political trouble. By January 1906, only fifteen months after the commissioners had triumphantly inaugurated their new wonder, the abolition of the RTC was being demanded by subway riders, by real estate developers, and, most important, by a new group of urban reformers: the progressives.

The Progressives and Rapid Transit

Progressivism swept across American politics and culture between the turn of the century and World War I. Appealing mainly to middle-class people of the old stock who feared that their traditional way of life was being endangered by the rise of big business and the growth of immigration, progressivism constituted a secular revival movement that sought the moral reawakening of the country. Progressivism was a wide-ranging, nationwide phenomenon that embraced an unusually large agenda, including settlement house workers who came to the aid of the immigrant residents of big city slums and urged their assimilation into the mainstream; good government campaign-ers who wanted to break old patterns of organized corruption and establish a new tradition of honesty unblemished by graft, patronage, and incompetence; and trustbusters who hoped to rein in the big corporations that dominated much of the economy.[38]

One of the reformers' leading causes was urban mass transit. Celebrated progressives such as Lincoln Steffens, Jacob Riis, Ida Tarbell, and Charles Beard wrote about mass transit; top reform journals such as *McClure's, Outlook,* and *Municipal Affairs* published articles on the subject. Urban mass transit received all this attention because the reformers believed it was an instrument for achieving two major goals: urban dispersal and anti-monopoly.

First, the reformers argued that improved transportation would enable immigrants to escape from their wretched urban slums. In his famous 1890 exposé of Manhattan's notorious Five Points, *How the Other Half Lives,* Jacob Riis showed that the poor lived in rundown, overcrowded tenements that lacked private toilets, hot and cold running water, and adequate heat. In 1890 the Lower East Side was perhaps the most crowded place in the world, with a population density of about 550,000 people per square mile; even more remarkably, most of the buildings on the Lower East Side were less than seven stories. Due to poor sanitation and extraordinary population densities, immigrant districts were ravaged by infectious diseases such as tuberculosis, typhoid fever, and diphtheria. The progressives also thought that these poor neighborhoods were breeding grounds of social pathology. In their view the Lower East Side, Hell's Kitchen, Greenwich Village, and Yorkville amounted to foreign colonies that encouraged newcomers to cling to their native languages, religions, and folkways rather than assimilating into American culture. This was dangerous. As hotbeds of crime, pauperism, and radicalism, these colonies threatened to eradicate traditional American social harmony and replace it with the kind of rigid, violence-ridden class structure that had already spread across Europe.

Rapid transit offered one solution. Although the movement of poor immigrants away from the center had been an unintended consequence of the Contract No. 1 subway, the progressives made urban dispersal a conscious priority. Through the construction of rapid transit lines radiating from the city core to the vast expanses of inexpensive land on the periphery, Jacob Riis, Adna F. Weber, John DeWitt Warner, and H. L. Cargill, among others, wanted to help immigrants settle in suburban towns embodying the small town ideal: clean, safe, prosperous, and conducive to voting, home ownership, and acculturation. There, families could blossom in wholesome surroundings and yet live close enough to the city for their breadwinners to commute to work downtown.[39]

Second, the reformers were alarmed that big transit corporations such as the Interborough Rapid Transit Company wielded monopoly power over vital urban transport. The problem in the case of the New York subway was the giveaway clause of the 1894 act, which had been added to get a bidder for the underground railway. Under the formula of municipal ownership and private operation, the Interborough enjoyed almost total freedom from government control and could refuse to build new subways with impunity. As Ray Stannard Baker put it, the Interborough had appropriated "the right to govern" New York City.[40]

The progressives blamed the Rapid Transit Commission for letting the IRT run amuck. The fact was that the reformers' conception of the public interest was very different from that of the merchants who led the Rapid Transit Commission. Indeed, the political culture of the merchant-commissioners was fundamentally different from that of the professionals and reformers who supplanted them; the two groups had dissimilar values, practices, and institutional relationships. Baker objected that the RTC took "a business engineering view, not a broad civic view" and that it adhered to "the old ideals of 'business interests' as compared with the 'public welfare.' "[41] Where the commissioners had sought to cooperate with the Interborough in building the Contract No. 1 subway and then in implementing its 1905 expansion plan, Baker chided them for breaching the inviolable wall that ought to separate the public and private spheres and for colluding with the corporation. One progressive, John DeWitt Warner, even accused the RTC of conspiring to promote Belmont's monopoly because so many of its new routes were simply extensions of the IRT.

Yet the commissioners did, in fact, possess a strong sense of public duty, as was clear from their broad, imperial vision of a magnificent subway that would enrich the entire metropolis. What distinguished the Rapid Transit Commission from its detractors, however, was that the merchants saw no conflict between the business and public interests and heartily approved partnerships between business and government. No commissioner, for example, ever censured August Belmont for exploiting his advance knowledge of the IRT's route by speculating in property on City Island in the Bronx; although terms such as "graft" and "waste" belonged in their vocabulary, the merchants would not have understood "conflict of interest." To the contrary, the commissioners prized August Belmont precisely because his business knowledge gave them a source of expertise that their understaffed, nonprofessional agency otherwise lacked.

This business formulation did not satisfy the progressives, however. Even before the Interborough-Metropolitan merger, they had taken the offensive against the RTC.

Before the subway merger took place, most progressives believed the Interborough Rapid Transit Company could be curbed by stimulating competition among transit companies. In 1903 the Citizens' Union drafted a measure, known as the Elsberg bill, that was designed to correct the monopolistic tendencies of the Rapid Transit Act of 1894 and prevent a single corporation from ever dominating rapid transit again. The Elsberg bill would separate the contracts for constructing new subways from the contracts for operating them, reduce the period of lease from fifty to twenty-five years, and allow the municipality to revoke future contracts more easily than could the so-called cast-iron Contract No. 1.

The Elsberg bill showed the limits of progressive reform. Although the progressives had moved beyond laissez-faire doctrine by articulating a new ideal of the public interest that legitimated the principle of business regulation, they remained suspicious of big government and preferred market solutions to active government intervention in the economy. The bill's position regarding municipal operation was a case in point. Even though the measure would authorize municipal operation of new subways for the first time, most reformers looked on this provision only as a tool for holding "a whiphand over the monopoly" during negotiations for more rapid transit construction. Only a fraction desired municipal operation for its own sake. Those like John DeWitt Warner who anticipated a city-run subway were not in the mainstream but belonged to a special current of the reform movement that spilled over into socialism.[42] Instead of sanctioning municipal operation, the bill relied on competition as its main control device. By lifting the legal constraints that stopped other transit companies from contending with the Interborough, the bill would produce a vigorous free market that would obviate the need for further encroachments on the private sphere.[43]

The Rapid Transit Commission opposed the Elsberg bill for repudiating its conciliatory approach with the companies. Blasting the reform initiative as "the most vicious piece of legislation I have ever seen," RTC counsel Albert B. Boardman complained that "it is proposed here to depart from every principle in the present Rapid Transit Act."[44] Furthermore, the RTC claimed that the Elsberg bill would disrupt its expansion plans. Recalling the long, arduous search for bidders for the first subway, the commissioners warned that imposing stiffer terms of contract would frighten away investors, seriously delaying further construction. "What this radical sentiment for municipal ownership and control of all public utilities may bring about . . ." Alexander E. Orr remarked, "I do not know. But that it is going to exert some influence upon the investment of private capital there seems little doubt. It is certain the city cannot do this work."[45] Why, Orr asked, should New York City break a tradition of public-private partnership that had succeeded so well with the first subway?

The RTC's staunch resistance prevented the bill from passing the state legislature when it was first introduced in 1903 and for the next several years. But the Interborough-Metropolitan merger on December 22, 1905, undermined public confidence in the Rapid Transit Commission, leading the state senate and assembly to approve the Elsberg bill in May 1906. Attitudes toward the RTC were changing so quickly, however, that even this amendment no longer seemed sufficient. There was little point in drumming up competition now that the Interborough had achieved a monopoly that mocked the very idea of free enterprise. More and more progressives in organizations such as the Citizens' Union, the City Club, and the People's

Institute grasped the futility of half-measures and now wanted to eliminate the RTC altogether and impose tougher restrictions on the IRT.[46]

The demand for stricter controls on public utilities became the main focus of the 1906 New York gubernatorial race. The Democratic candidate, wealthy newspaper publisher and maverick politician William Randolph Hearst, did not share mainstream progressivism's reverence for property rights. A brash urban populist who denounced the corporate plutocracy in his campaign speeches, Hearst advocated municipal ownership (and operation) of gas, electric, and transit companies. The Republican standard-bearer, Charles Evans Hughes, a prominent attorney and former Cornell University law professor, won the gubernatorial nomination by virtue of his recent investigations of financial shenanigans in the life insurance industry and the local gas utility. Steering a middle course between the corporate boardroom and the socialist fringe of the reform movement, Hughes favored administrative regulation of the public utilities. Although Hearst's vigorous campaign touched a raw nerve, Hughes, who had an edge in the Republican-leaning New York State, squeaked through with a plurality of 58,000 votes out of the nearly 1.5 million cast.[47]

Shortly after entering office, Governor Hughes sponsored legislation, called the Page-Merritt bill, to create a pair of public utility regulatory commissions, one for New York City and the other for upstate. These commissions would be extremely strong agencies with broad jurisdiction over electric, gas, railroad, and mass transit corporations; with authority to conduct investigations, set rates, and order schedule changes; and with large professional staffs as needed to enforce their decisions. In the minds of its progressive sponsors, the Page-Merritt bill aimed at altering New York's political culture. The bill would confirm that the relationship between government and private utilities was formal and adversarial rather than private and cooperative; and it would establish strong government agencies with the resources to carry out the "public interest."

Many businessmen opposed this reform and tried to derail it through lobbying, bribery, and other means. According to some newspaper accounts, agents of the Interborough Rapid Transit Company stole an early draft of the governor's proposal and handed it to August Belmont even before it was seen by many assemblymen and senators.[48] But Governor Hughes, appealing to Republican legislators in the name of party loyalty, kept enough of his supporters in the fold to defeat this conservative challenge. The Page-Merritt bill passed the GOP-dominated legislature easily, and Governor Hughes signed it into law on June 6, 1907.[49] Three weeks later, on the first day of July, the new Public Service Commission for the First District (PSC) replaced the old Board of Rapid Transit Railroad Commissioners.

Despite their failings, the Rapid Transit Commissioners had expressed a bold vision of underground rapid transit as a public enterprise that was essential to the metropolitan economy. They overcame the daunting political and geographical obstacles that had initially blocked their project and completed a splendid rapid transit railway that was widely regarded as the world's best. These were substantial achievements. Time would tell whether the merchant-commissioners' successors would accomplish as much and what the consequences would be of the new political system of more open conflict between business and government.

PART II

THE POLITICIANS AND THE SUBWAY

6

THE DUAL CONTRACTS

Congestion

Never before had so many people been packed into Manhattan as in the first decade of the twentieth century. The population of Manhattan, which had exploded during the nineteenth century, peaked in 1910 at 2,331,542. In that year Manhattan ranked among the most congested urban places in the world. More people resided on this small, twenty-three-square-mile island than in thirty-three of the nation's forty-six states.

Manhattan was vastly more populous than the other four boroughs. In 1910 forty-nine of every one hundred New Yorkers lived in Manhattan. At a time when many parts of the Bronx, Brooklyn, Queens, and Staten Island were thinly settled, Manhattan bristled with tenements, row houses, apartment buildings, and factories. Because the island contained only 14 percent of the city's total land area, it was exceptionally crowded. Manhattan had an average density of 161 people per acre in 1910, far above Brooklyn (32.5 residents per acre), the Bronx (15.6), Queens (3.8), and Staten Island (2.2). And some Manhattan neighborhoods reported densities much higher than the average. One-sixth of all New Yorkers lived below Fourteenth Street, on one eighty-second of the city's land area. The single most crowded district was the Lower East Side, where densities exceeded 700 people per acre in 1910.[1]

Many progressive reformers were alarmed by the tremendous congestion and sought to move residents out of Manhattan. The best way to do so was through public works that tied the island to the rest of the city. The Williamsburg Bridge opened in 1903, and many Jewish immigrants relocated from the jam-packed Lower East Side to the bigger, more comfortable flats available directly across the East River in the Williamsburg section of Brooklyn. So many did so that the *New York Tribune* dubbed the new span "Jews' Highway." Six years later the Manhattan and Queensboro bridges were completed, prompting still more people to leave Manhattan.

But these bridges extended only from one riverfront to the other and did not reach into the interior of the boroughs. Thousands of acres remained relatively undeveloped in eastern and southern Brooklyn, the Bronx, Queens, and Staten Island. The IRT had begun to shift the population from Manhattan to the Bronx, but to resettle Manhattanites in these more distant outlying areas, new subways had to be built.[2]

Triborough

Six months after it was established, the Public Service Commission completed plans for a new subway, the Triborough system. This proposal for a $150 million subway was designed to implement the reform agenda of destroying the Interborough monopoly and relocating poor Manhattanites to the outskirts.

The Triborough would comprise three major routes in Manhattan, the Bronx, and Brooklyn, and would total 144 single-track miles, nearly double the length of the IRT subway. (See Map 2, Triborough Subway Plan, 1908.) One route, the Broadway-Lexington, would go from the Battery up the east side of Manhattan to 138th Street in the Bronx, where it would split in two, with one branch following Mott Avenue and Jerome Avenue to Woodlawn Cemetery, and the other going along Southern Boulevard and Westchester Avenue to Pelham Bay Park. A second route, the bridge loop, would relieve the horrible

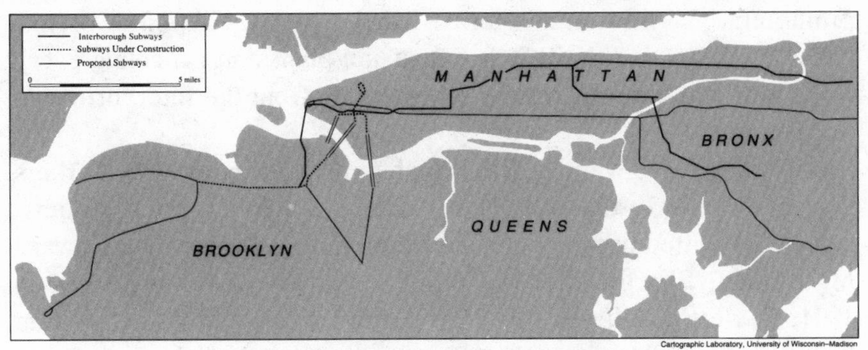

2. Triborough Subway Plan, 1908.

traffic congestion that slowed bridge and ferry travel across the East River, New York City's worst bottleneck and the busiest water corridor in the world. This bridge loop subway would circle through parts of Manhattan and Brooklyn, running from Canal Street up to Delancey Street, across the Williamsburg Bridge, through Brooklyn via Broadway, Lafayette Avenue, and Flatbush Avenue, over the Manhattan Bridge (which was under construction), and then back to its starting point at Broadway and Canal. The third route, the Fourth Avenue, would sweep south from Brooklyn's business district through the rural flats of Sunset Park and Bay Ridge to the shores of the Narrows.[3]

These three routes were originally laid out by the old Board of Rapid Transit Railroad Commissioners, not by the Public Service Commission. Indeed, the RTC had let the contracts for the construction of the bridge loop in June 1907, and crews of workers were already excavating it by the time the PSC took office. But even though the PSC inherited these lines from its predecessor, Triborough nonetheless embodied an important new direction in rapid transit planning. The five public service commissioners were all attuned to progressivism's broad perspective on rapid transit.[4] Rejecting the RTC's discredited approach of trying to cooperate with the Interborough in building more subways, the PSC conceived of Triborough as an independent railway that would compete with the IRT and—in the phrase of the *New York World*—"protect against the abuses of private management and the greed of thumble-rigging promoters."[5] The commissioners carried out their policy preferences by reviewing the RTC's expansion plans and canceling every route that could not be operated separately from the Interborough's underground.[6]

The PSC also wanted Triborough to disperse poor New Yorkers from the center to the outlying districts. No public service commissioner advocated urban deconcentration more strongly than Edward M. Bassett. Shortly after the original subway was inaugurated in 1904, Bassett concluded that New York City's rapid transit system had two glaring flaws. The first was the paucity of rapid transit facilities between Brooklyn and Manhattan.[7] Before the Contract No. 2 subway or the Manhattan Bridge opened, and the Brooklyn and Williamsburg bridges remained the only direct connections across the East River, Bassett believed that more transriver subways were needed to ease the traffic between these two bustling urban centers and to spur the growth of remote sections of Brooklyn such as Bay Ridge, Flatbush, and Canarsie.

The second flaw was the IRT's spectacular impact on new neighborhoods such as the Upper West Side, Harlem, and the Bronx. Bassett considered these areas ideal building sites for one-and two-family homes, which would preserve much of the rural landscape and nurture strong communities. But the pent-up demand for more living space was so great in New York City that when the subway entered a new territory, the price of real estate skyrocketed so that only multistory apartments or unsightly tenements could be built. Bassett complained that this kind of unrestricted, high-density growth blighted the city. According to Bassett, the problem was that not enough rapid transit lines were being built at one time. With only one or two subway routes being completed at each stage, the amount of land opened for development remained woefully inadequate, resulting in high densities of population that overwhelmed the new districts. In addition to designing through-running routes that would go below the East River, Bassett also wanted to construct a large number of subways.

Bassett claimed that both of these flaws stemmed from the same source: New York City's daunting geography. Most of the city's rapid transit lines stopped at the water's edge in "stub-end terminals''—such as South Ferry in Manhattan and Atlantic Avenue, Fulton Street, and Broadway in Brooklyn—that compelled a rider who wanted to continue his journey to leave the train, board a ferry, and catch another train on the other side of the river. Bassett blamed these stub-end terminals for causing the severe traffic jams that clogged the waterfront during the morning and evening rush hours and for forcing commuters to reside close to their jobs downtown rather than move farther out of town. He argued that the new subways should be designed as pendulums, swinging from the outskirts of the city, through the business district, and then across the river, without stopping at the waterfront. To Bassett, who was honored many years later as a pioneer of the American city planning profession and who is perhaps best remembered as co-author of New York City's 1916 comprehensive zoning ordinance, good subway design provided a means of rationalizing land development without encroaching too much on the private market. If enough pendulum subways were built, New York would be transformed into a ''round city,'' with so much open space available in new territories that real estate prices would stay low and there would be a pattern of balanced, low-density growth.

To be sure, Triborough did not completely satisfy the reformers' aim of dispersing the population. The problem was that the progressives' goal of decentralizing the city conflicted with their goal of competing with the Interborough. The Public Service Commission could not push its lines very far beyond the edge of settlement because its independent subway might go bankrupt if it did not carry enough traffic. Consequently, relatively little mileage was slated for new, unpopulated areas, and Triborough probably would not realize Bassett's vision of a round city of well-controlled growth.[8] To many progressives this was a question of priorities. As long as August Belmont exercised veto power over rapid transit construction, no new subways would be built anyway, and the poor would remain trapped in Manhattan.[9]

The City Club, the Reform Club, the People's Institute, and other progressive organizations endorsed Triborough. Local businessmen who represented neighborhoods that lay along its routes—such as the East Twenty-third Street Association, the South Brooklyn Board of Trade, the Fourth Avenue Subway League, and the North Side Board of Trade—also boosted it.[10]

Downtown businessmen were far more critical, however. The strongest opposition came from the Chamber of Commerce of the State of New York, which appointed a special committee to investigate the Triborough proposal. Headed by merchant and manufacturer Eugenius H. Outerbridge, the committee concluded that Triborough was only the latest in a series of progressive blunders—such as the passage of the Elsberg bill and the creation of the Public Service Commission itself—that had undermined the Chamber's great achievements in founding New York City's first subway.[11] According to the Outerbridge committee, Triborough would entail higher capital and operating charges than an expansion of the IRT, which would benefit from scale economies. Triborough's routes were poorly designed, too. For instance, the Public Service Commission's Broadway-Lexington route would run within a few blocks of the IRT subway south of Grand Central Terminal, on Manhattan's east side, duplicating an existing underground when there was no subway service at all below Times Square on the island's west side. This lack of subway service on the west side below Forty-second Street was particularly glaring because the Pennsylvania Railroad was then completing its magnificent new station at Seventh Avenue between Thirty-first and Thirty-third streets; without subway connections, Penn Station would be marooned and passengers would have a tough time getting to it. Too much trackage, moreover, had been laid out in the rural, thinly settled sections of the Bronx and Brooklyn where traffic levels would be too low to support rapid transit.[12] Noting Triborough's emphasis on serving these outlying districts and on competing with the Interborough, the Outerbridge committee complained that the plan advocated business inefficiency in the name of public service. Accusing the PSC of "proceed[ing] in violation of the fundamental economic laws which would govern private enterprise," the committee warned that Triborough would produce an operating deficit of $100,000 to $235,000 annually.[13]

For the reformers, however, subway building involved far more than just dollars and cents. Their support for Triborough reflected two broad issues that were at the heart of the progressive movement: monopoly capitalism and urban political corruption. Progressive journals such as *Collier's, American* magazine, *Outlook,* and particularly *McClure's* were publishing sensational exposés of the tyranny of corporations and political machines. In 1902–3, Lincoln Steffens wrote a series of articles called "The Shame of the Cities" for *McClure's* that revealed big city governments and powerful local businessmen were ensnared in a web of graft; examining New York, Philadelphia, Pittsburgh, and St. Louis, Steffens showed how corrupt public officials and dishonest businessmen conspired to arrange sweetheart deals for street-paving contracts and railway franchises, to steal elections, and to protect prostitution, gambling, liquor sales, and other vices. At the same time *McClure's* serialized another famous exposé, Ida M. Tarbell's history of the Standard Oil Company, which disclosed how John D. Rockefeller had used ruthless business practices to seize control of the oil industry. As a result of muckrakers such as Steffens and Tarbell, the political machine and the corporation became symbols of the greed and graft that were thought to be sabotaging traditional American values.

New York City's progressives stressed these themes, too. Reform organizations such as the City Club and Citizens' Union and the journal *Municipal Affairs* pounded Tammany Hall for selling judgeships, extorting bribes from contractors who wanted building permits, and pocketing kickbacks from brothels and pool halls. Similarly, the reformers criticized the monopoly power and exorbitant profits of giant corporations such as August Belmont's Interborough, Third Avenue Railroad Company, New York Realty Company, and Consolidated Gas.[14]

For most progressives the Triborough plan provided a means of undermining the Interborough monopoly and challenging August Belmont's cozy relationship with Tammany Hall. The public service commissioners thus defended Triborough against the conservative onslaught led by the Chamber of Commerce. But they nonetheless heeded the Outerbridge committee's criticism that progressive reforms like the Elsberg Act of 1907 poisoned the atmosphere between business and government. The PSC was committed to a market solution and wanted Triborough to be financed, built, and operated by business, not by government. Without new contractual arrangements to reduce investors' risk, however, the commissioners feared that financiers would not bid on Triborough.[15]

The Public Service Commission devised several fiscal reforms in order to promote private investment in rapid transit. One proposal would increase the length of time for operating a subway built with private capital from the twenty years fixed by the Elsberg Act to any period approved by the authorities. The second consisted of a new mechanism, the indeterminate franchise, for regulating the private operator. Under the terms of the indeterminate franchise, the municipality would award a long-term franchise to a company that invested its own money in rapid transit construction. In order to guarantee that the leaseholder did not control the Triborough subways as tightly as the Interborough Rapid Transit Company did, the indeterminate franchise would enable the city to acquire the subway ten years after the start of operation by compensating the company for its outlays. The sum of this payment would be equal to the total cost of the company's investment plus 15 percent at the outset; as the value of the subway depreciated over time, the city's payment would decrease. According to Commissioner Milo R. Maltbie, the indeterminate franchise would give the company the security of a long franchise at the same time that it provided the city with an escape clause in case the company misbehaved.[16]

Governor Charles Evans Hughes endorsed these PSC measures in January 1909. Legislative approval followed quickly, and they were enacted by May 1909.[17]

The next step was for the PSC to request permission from the Board of Estimate and Apportionment to draft contracts for the Triborough plan. As restructured during the consolidation of Greater New York in 1898, the Board of Estimate was a powerful executive body that occupied "the center of gravity in the city's political process," exercising immense authority over law-making, the budget, and franchises.[18] A 1900 law gave the Board of Estimate the power to appropriate municipal funds for rapid transit, and five years later it acquired the right to approve the location of subway routes, too.[19]

Previously, the Board of Estimate was dominated by Tammany Democrats, such as Mayor George B. McClellan and Comptroller

Herman Metz, who detested the Republican-created, Republican-led New York State Public Service Commission and were close to the Interborough Rapid Transit Company. Under the influence of McClellan and Metz, the Board of Estimate had blocked earlier PSC requests to assign contracts for the construction of parts of the Fourth Avenue subway. Consequently, prospects for Triborough's approval did not look very bright. But 1909 was an election year, and many progressives allied with the Republicans behind "fusion" candidates in an effort to form a united front against the Democratic machine.[20] These fusionists were generally eager to cooperate with the Public Service Commission, and most of them accepted the commission's policy of building Triborough with private capital. In November, Republican-fusionists captured six of the eight positions on the Board of Estimate, including the comptroller's office, the president of the Board of Aldermen, and the presidencies of four of the five boroughs.[21]

The reformers failed, however, to capture New York City's top prize, the mayoralty. A split in the progressive ranks between the fusionist candidate, a colorless banker named Otto T. Bannard, and the more extreme Civic Alliance nominee, flamboyant newspaper publisher William Randolph Hearst, opened the door for the Tammany Hall standard-bearer, State Supreme Court Justice William J. Gaynor of Brooklyn. Unlike many Tammany nominees, Gaynor enjoyed an unblemished reputation for political independence and moral integrity. Realizing that an economic recession and recent revelations of municipal corruption made 1909 a good year for reformers, Tammany boss Charles Murphy went with Gaynor because he thought a more regular Democrat candidate could not win. During the campaign Gaynor passionately denounced the traction companies for being greedy and for corrupting city politics, yet he warily refused to either declare his support for Triborough or renounce cooperation with the Interborough Rapid Transit Company in building new subways. The result was that when Gaynor won the election with 43 percent of the vote, no one could tell what he might do.[22]

Nonetheless, the fusionist victory meant that the Public Service Commission could count on a sure majority of the Board of Estimate. "There is now in office . . ." PSC Chairman William R. Willcox declared, "a Board of Estimate whose members are pledged to rapid transit progress and who have indicated their readiness to devote all of the money that can be spared to rapid transit progress."[23] In January 1910, Board of Estimate and Public Service Commission officials began to confer about contracts for Triborough.[24] The only surprise occurred when Mayor Gaynor abruptly announced his support for building the cheaper IRT extensions rather than the more extravagant Triborough. A bad-tempered, sharp-tongued man, Gaynor frequently made sudden, baffling political changes. He became so famous for his erratic behavior that his biographer described him as an "intransigent individualist" whose career "was a long record of almost startling variations from type."[25]

But Gaynor's change of heart did not affect the outcome. With the Republican-fusionists commanding a majority of the votes, the Board of Estimate approved the Triborough contracts in August 1910. On September 1, 1910, the PSC advertised one set of contracts for the construction, equipment, and operation of Triborough with private capital under the indeterminate franchise formula and a second set of contracts for the construction of particular Triborough sections with municipal funds. The second set was intended as an insurance policy in the event financiers did not respond to this invitation, and the PSC also offered a second set of contracts for the construction of particular Triborough sections with municipal funds.

On October 20, 1910, the PSC opened the first set of bids, only to discover that not a single company submitted a proposal. This was a staggering setback for the commission after its three years of work on the Triborough plan. Seven days later the commissioners were relieved to find that twenty-three offers had been made to build Triborough with municipal funds. Gratified, the PSC then began to weigh the possibility of municipal construction. But businessmen from the Chamber of Commerce and the Merchants' Association denounced the commissioners for even considering this extreme step.[26] The Outerbridge committee admonished the PSC, saying that the "folly" of municipal construction was "fraught with grave dangers to the success of the undertaking and the credit of the City of New York."[27]

Chairman William R. Willcox did not know what to do. A weak

and indecisive leader, Willcox was caught between the conservative businessmen who abhorred public construction and the reformers who favored it as a means of saving Triborough. Unable to make up his mind, Willcox vacillated, and the PSC floundered.

Fortunately, a shrewd businessman soon seized his opportunity.[28]

William G. McAdoo

On November 18, 1910, President William G. McAdoo of the Hudson and Manhattan Railroad Company (H&M) sent the PSC a bold plan for a new subway covering much of Manhattan and reaching into the Bronx and Brooklyn. "The one great big thing," McAdoo proclaimed, "is to get some competition into the New York transit situation. The city will never get the benefit of the best development of the rapid transit service until there is competition in operation."[29]

In addition to contending directly with the Interborough, this new subway would be extremely important in its own right. McAdoo had completed two subway tunnels, then called the Hudson tubes and now known as the Port Authority Trans-Hudson, or PATH, under the Hudson River between New York and New Jersey. By linking the Hudson tubes to his projected New York subway, McAdoo proposed to create a regional transit system of remarkable scope.

A Tennessee lawyer and businessman, William G. McAdoo had tried to consolidate Knoxville's street railways in 1889. Three years later his company went bankrupt, and he moved to New York City to start over, as a corporate lawyer on Wall Street.[30] He soon recovered financially. In 1901, McAdoo tackled one of New York City's most vexing engineering projects: a tunnel under the Hudson River that had been started in 1874 and remained unfinished three decades later. This tunnel was the brainchild of a western railroad engineer who had the perfect New York City name of DeWitt Clinton Haskin. Haskin thought a tunnel was needed to provide direct rail access across the Hudson River. Since the Lackawanna, Central of New Jersey, Pennsylvania, and Erie railroads terminated on the Jersey side of the river, passengers had to take a ferry to New York City. A ferry ride ordinarily lasted no more than ten or fifteen minutes, but the boats were often delayed by rain, snow, or fog and sometimes collided with lighters, steamers, barges, or other vessels. To solve this problem, Haskin proposed drilling a pair of single-track tunnels from the Jersey City–Hoboken border to a large union station that would be located under Washington Square in Greenwich Village. Haskin succeeded in completing about 40 percent of the 4,800-foot-long northern tunnel, but chronic financial and engineering difficulties brought the project to a halt in 1882. Although the British contracting firm of S. Pearson & Son resumed the work seven years later, it soon ran out of money in the depression of the 1890s, and the tunneling stopped again.

Even though the tubes had long since become a notorious fiasco, William G. McAdoo sensed that the demand for transriver transport had grown dramatically, so he greatly expanded Haskin's scheme into a regional rapid transit network that would connect Manhattan to northern New Jersey's busiest railroad stations and largest cities. McAdoo's company, the Hudson and Manhattan, finished Haskin's original tunnel from the Delaware, Lackawanna, and Western Railroad's Hoboken terminal to Greenwich Village, and added an extension up Sixth Avenue to Thirty-third Street in Herald Square. Along with this uptown route, the H&M also dug a downtown route that ran from Cortlandt Street in lower Manhattan, under the Hudson River, through Jersey City, and on to Newark.[31]

That was only the beginning. As McAdoo explained in his prospectus of November 15, 1910, he dreamed of creating a second subway network that would join with his Hudson tubes and fuse New York and New Jersey into a single metropolitan region. After this $150 million system opened, city residents would be able to ride from Brooklyn Heights to a railroad station in Jersey City or Hoboken in only forty-five minutes without ever seeing the East or Hudson rivers. McAdoo's new network would consist of three routes totaling fifty-five miles of single track. The main line, a modification of Triborough's Broadway-Lexington subway, would go up the east side from the Battery to River Avenue in the southern Bronx. The second route would follow Broadway from Greenwich Village (where it would connect with the Broadway-Lexington, at Tenth Street) to Thirty-third Street (where it would connect with the H&M). The third subway would begin at Church Street in lower Manhattan (with a link to the H&M's downtown tube), pass under Wall Street, and cross the East River to Brooklyn.[32]

McAdoo's proposal delighted the Public Service commissioners, who felt that their goal of stimulating competition with the Interborough was now within reach. "One thing is apparent," William R. Willcox declared, "and that is that a responsible company is ready to undertake the operation of the tri-borough system with minor modifications." The *New York Evening Post* spoke for many reformers in saying that the McAdoo offer signaled the death of "monopoly with the Interborough type of management—stupid, incredibly short-sighted, grasping, and now out-manoeuvered"[33]

August Belmont had other ideas, however. The IRT chairman had never worried that the Triborough might compete with his subway; he thought that the PSC's system was so badly designed and so uneconomical that no transit company would want to operate it. For this reason the Interborough chairman reassured Lord Rothschild in February 1909 that "any talk of building new subways or competing lines by rival or outside interests has died out, and it is becoming evident that no one can possibly give relief outside of our corporation."[34]

Belmont's low regard for Triborough determined his strategy for negotiating with the Public Service Commission. Not long after the PSC took office, IRT president T. P. Shonts began to confer with Chairman William R. Willcox about the possible extensions to the IRT's subways and elevateds, but although the Interborough submitted a series of proposals to the PSC, Belmont and Shonts probably never wanted to come to terms. Rather, their overtures were designed to create the impression that the IRT was bargaining in good faith; in reality they contained routes and financial conditions that were so unattractive, the PSC was almost certain to reject them.[35] In May 1909, Belmont told Gardiner M. Lane, a Boston banker and IRT director, why the company bothered to submit a plan that stood no chance of being accepted: "There is no danger of competition in the building of subways at the present time, and I really do not entertain the conviction that whatever we propose will be accepted at the moment, but we would be putting ourselves on record in the line of progression, which is important.''[36] Belmont intended to wait until Triborough's failure put the PSC under great pressure to make a deal that would bring about new subways. The IRT would then present a proposal containing stiff terms, including the use of municipal capital to finance construction.[37] ''If we handle it properly,'' Belmont confided to Lane, ''we will ultimately have the use of the city's credit, as we had before.''[38]

Although Belmont accurately predicted the financiers' rejection of Triborough, he did not anticipate McAdoo's move. Belmont took the H&M's November 1910 proposal very seriously; the Interborough faced a real threat of competition for the first time since the Metropolitan merger of December 1905, so Belmont decided to knock the H&M out of the running. On December 5, T. P. Shonts submitted the IRT's plan to the PSC. It included two new subways that would connect with the original Contract No. 1 line. One would go down the west side from the IRT's Times Square stop to the Battery and cross the East River to Brooklyn. The other route would go up Lexington Avenue from Grand Central Terminal to the Bronx. When joined with the old subways, these new IRT undergrounds would take the shape of an H, with the letter's crossbar formed by the stretch of the Contract No. 1 route that ran between Times Square and Grand Central on Forty-second Street. In addition to these subways, the Interborough wanted to erect more third-tracks to improve the express service on some of its elevateds, lengthen several els in the Bronx, and send the Second Avenue el across the newly opened Queensboro Bridge.[39]

As Belmont had hoped, this Interborough overture checked the Hudson and Manhattan. Speaking at a City Club luncheon a few days before the IRT submitted its proposal, McAdoo vented his frustrations at trying to vie with the Interborough:

> For the last five years, with all the powers of the Public Service Commission, with all the powers of the city, and with all the powers of the municipality, you have never been able to get a single bit of new subway constructed to relieve the present demand of the community [for lines to the outskirts], and it will always be so as long as one company has complete control of the situation.[40]

According to his biographer, John J. Broaslame, William G. McAdoo faced "a frustrating dilemma" in trying to implement his PSC offer because "the finances of the Hudson and Manhattan may well have been stretched to the breaking point in his initial proposal."[41] Too weak to maintain a substantial financial commitment indefinitely, McAdoo gave the PSC only one month in which to approve his plan. The Interborough's December 5 proposal further impaired the H&M's ability to raise money by frightening away investors who did not want to compete with the Belmont monopoly. By this time, moreover, Belmont had established a close relationship with J. P. Morgan, Jr., whose House of Morgan planned to underwrite the Interborough's bonds for new subways. Morgan's support of the IRT further weakened McAdoo's position on Wall Street. As a result, McAdoo withdrew his proposal when the time limit expired on December 15, 1910.[42]

But the Interborough had finally been flushed from cover. "The stupid, hitherto immovable Interborough has found itself compelled by the mere threat of competition to submit an offer to build and operate new subways which it ought to have . . . made long ago," the *New York Evening Post* proclaimed.[43] William R. Willcox expressed satisfaction that the Interborough had finally submitted a serious proposal and pointed to "several attractive features" in the IRT plan.[44] The commissioners felt more conciliatory toward the IRT now that the collapse of the Triborough and H&M plans narrowed their options, but they were concerned that accepting this IRT proposal would mean the end of their goals of urban dispersal and trust-busting. Because most of the Interborough routes were located in Manhattan and the Bronx, some commissioners thought the development of outer Queens and Brooklyn would be retarded. Others suspected that the IRT offer was designed to protect the company's monopoly. In his letter to Willcox, T. P. Shonts suggested that the IRT pay $75 million of the subway proposal's total cost of $128 million and that the municipality contribute $53 million. The problem was that this IRT request for city funds would have exhausted virtually all the funds available for rapid transit, preventing the PSC from signing a contract with another company that might compete with the Interborough.[45]

Despite these misgivings, the commissioners decided on December 20, 1910, to forward the IRT plan to the Board of Estimate for inspection. Afraid of taking a firm stand one way or the other, the Public Service Commission was simply passing the buck to the Board of Estimate. If the Board accepted Shonts' proposal, then the PSC would negotiate with the Interborough; if the Board preferred the Triborough alternative, then the PSC would draft contracts for its construction with municipal capital.

But the Board of Estimate could not reach a decision, either. With Mayor William J. Gaynor supporting the IRT, Board of Aldermen President John Purroy Mitchel and Comptroller William A. Prendergast holding out for private construction of Triborough, and four of the five borough presidents refusing to take a stand, the Board of Estimate remained deadlocked through December and into January.[46]

George McAneny and the Dual System

The stalemate broke on January 10, 1911, when President Edwin W. Winter of the Brooklyn Rapid Transit Company (BRT) issued a subway plan. A giant corporation that controlled nearly all the elevated and street railway lines in the borough of homes and churches, as Brooklyn was known, the BRT would be a formidable rival for the IRT.

Winters designed his proposal to solve two problems that bedeviled his corporation; its need to put the jumbled transit lines in better order and its need to create a distribution system in Manhattan. The BRT's routes were as tangled as its corporate history. Formed as a securities holding company in 1896 and greatly enlarged through a series of mergers and leases, the BRT was a big, loosely knit corporation that had swallowed such independent elevated companies as Brooklyn Union, Kings County, and Seaside and Brooklyn Bridge; old steam surface railroads such as Culver, Sea Beach, Brighton, and West End, which had been built in the 1860s and 1870s to take passengers to Brooklyn's Atlantic Ocean beaches; and a number of electric and horse railways. Never combined into a coherent network, these lines remained a confusing, overlapping admixture.

Even more troubling was the fact that the BRT could not get its passengers to Manhattan. Many BRT riders commuted to downtown Manhattan, but the company's elevateds ended at ferry terminals on the Brooklyn side of the river; for years there was only one BRT line that touched Manhattan—a short el that shuttled across the Brooklyn Bridge, over which many Myrtle Avenue and Fulton Street trains passed to Park Row.[47]

Winters wanted to take over the PSC's Fourth Avenue and bridge loop subways. Both of these routes had been laid out by the Rapid Transit Commission, and both were now under construction by the Public Service Commission. The Brooklyn Rapid Transit Company would connect the Fourth Avenue to its south Brooklyn railways, which included the Third Avenue elevated as well as the old steam railroads to the Atlantic Ocean beach resorts. The BRT would convert these railroads to rapid transit service and modify them to feed trains into the Fourth Avenue. Sea Beach or Culver trains would switch onto the Fourth Avenue tracks at the Fortieth Street junction and roll through downtown Brooklyn, then enter the bridge loop and head across Manhattan Bridge.[48]

This BRT offer angered August Belmont, who resented "the efforts of the Brooklyn Rapid Transit Co. to force their way into it and secure something for themselves. . . . Our proposal," Belmont insisted, "is by far the best for the city."[49]

Hardly anyone agreed with him. Most reformers perceived Winters' offer in the tradition of the Triborough and H&M plans and viewed the Brooklyn Rapid Transit Company as the latest alternative to the Interborough. But a few began to consider the possibility of integrating the IRT's and BRT's separate proposals. Most of the credit for this idea belongs to George McAneny, who was elected president of the borough of Manhattan in the 1909 reform sweep. A native of Greenville, New Jersey, McAneny graduated from Jersey City High School in 1885 and took a job as a reporter for the *New York World*. As a young newspaperman McAneny entered the genteel circles of the reformers who were fighting to uproot Tammany corruption and instill a new spirit of public service distinguished by honesty, economy, and professionalism. With his deep faith in rationality and his conviction that people would act reasonably if given the facts, McAneny thought that disinterested experts could reconcile divergent points of view and point the way to a common good that would transcend parochial special interests. In 1892, McAneny left journalism for a position as assistant secretary of the National Civil Service Reform League. He became an acolyte of the great Republican reformer Carl Schurz, and he spent the rest of his life trying to create an expert, nonpartisan bureaucracy. In 1902, McAneny assisted in drafting a civil service code for the City of New York, and five years later he helped incorporate the Bureau of Municipal Research in an attempt to systemize government.

McAneny first became interested in rapid transit after his election as president of the progressive City Club in 1906. Unlike many progressives, who struck a fairly even balance between the objectives of urban dispersal and anti-monopoly, George McAneny was more concerned about using rapid transit to stimulate city development than to curb the IRT. He became active in the new profession of city planning that arose from the progressive movement before World War I and that was also shaped by such influential precedents as Daniel H. Burnham's 1909 *Plan for Chicago*. Progressivism and city planning were closely intertwined, sharing such common concerns as housing reform, residential overcrowding, and rapid transit improvements. With planners such as George McAneny (and his friend Edward M. Bassett of the Public Service Commission), however, the emphasis began to shift from the social movement's impassioned crusade against particular problems—model tenements, park design, the layout of streets and boulevards—to the profession's cooler, more broad-based concern with the larger urban environment. Thus, McAneny was as interested in remaking the city as in relocating the poor to the outskirts. McAneny understood city planning "as the science of directing the actual building and growth of a city along proper and rational lines."[50] He viewed rapid transit as an instrument for replacing the squalid nineteenth-century metropolis with an efficient city that would be attractive, prosperous, healthy, and orderly. In this ideal city the population would be spread to outlying sections; residential neighborhoods segregated from business and manufacturing districts; heights and sizes of buildings regulated to prevent overdevelopment; streets arranged logically; traffic congestion eliminated; and parks, playgrounds, and public baths supplied. Because this new, sparsely settled city would be spread across a huge expanse, rapid transit would be needed to tie its distant sections together and to avoid traffic congestion.[51]

McAneny also had distinctive ideas about the relationship between business and government. Unlike many New York reformers, McAneny drew no moral distinction between the private and public spheres and never indulged himself in emotional anti-business rhetoric. To the contrary, he was an example of historian James Weinstein's model of the "corporate liberal" reformer who united business with government and called for positive state action to correct economic and social problems. A small man with a slight, almost frail build, a thin, handsome face, and a neatly trimmed Van Dyke beard, George McAneny was a skilled administrator who had a reputation for great tenacity. With his low-key personality, patience, and good manners, McAneny got along very well with people and moved easily from group to group. During his long career McAneny shuttled from elected office (president of the borough of Manhattan, 1910–13; president of the Board of Aldermen, 1914–16; comptroller of New York City, 1933) to "good government" organizations (City Club, Regional Plan Association, and Citizens Budget Commission), to businesses (executive manager of the *New York Times,* 1916–21; president and then chairman of the board of trustees of the Title Guaranty Trust, 1934–46), to semi-public corporations (president and later chairman of the 1939–40 World's Fair). McAneny knew how the city's multilayered bureaucracy operated, and he had personal access to important leaders through his web of contacts and acquaintances. With his excellent political and business connections and his belief in rational action and social responsibility, McAneny represented a new breed of dealmaker that arose in early twentieth-century New York. As New York evolved into a gigantic metropolis where no single group could exert its will anymore and where a large number of business elites competed for influence, dealmakers such as McAneny (and, in later periods, Samuel Untermyer and Richard Ravitch) facilitated decisions by forging links among top corporate and government leaders.[52] Significantly, a journalist once praised McAneny as a peacemaker "who could bring disputants together and work out an amicable solution."[53] Men like George McAneny provided the glue that kept this bewilderingly complex city from falling apart completely.[54]

George McAneny was the right man to break the subway planning deadlock. He sensed that the negotiating stalemate could be ended and the reformers' objectives of breaking the Interborough's power and stimulating urban deconcentration could be achieved by combining the IRT and BRT proposals. By signing contracts with both the Interborough Rapid Transit Company and the Brooklyn Rapid Transit Company, the City of New York could establish an economic rivalry that would let it play one company against the other instead of being beholden to the Interborough. Yet McAneny himself was less concerned about competition than about taking the best of both propositions and putting them together in a way that would benefit the whole city. To McAneny, coordination was always more important than competition. And with the IRT focusing on Manhattan and the Bronx, and the BRT concentrating on Brooklyn, the plans complemented each other nicely. The city could thus create two huge systems, each complete in itself, that would multiply its total rapid transit mileage, particularly in the outlying districts.

On January 19, 1911, the Board of Estimate and Apportionment created a three-member transit committee that would confer with the Public Service Commission, the Brooklyn Rapid Transit Company, and the Interborough Rapid Transit Company and then recommend policies for the new subways. Three days later Mayor Gaynor named McAneny, as Manhattan borough president, chairman of this committee; the other two members, Bronx Borough President Cyrus C. Miller and Richmond Borough President George Cromwell, deferred to McAneny and had relatively little impact on the committee's deliberations.[55] As transit historian Peter Derrick noted, the formation of this committee marked "a turning point" in the long, exasperating drive for more subways.[56] The Board of Estimate soon replaced the Public Service Commission as the source of major transit decisions, officials dropped the idea of building an independent railway like Triborough once and for all, and McAneny's committee attempted to enlist the IRT and the BRT in a joint construction program.

McAneny's committee quickly reached an understanding with

the PSC on the subway routes and financial terms to be obtained from both companies. It also received better subway plans from the two companies. The Interborough agreed to convert an old street railway tunnel, the Steinway tunnel that went from Forty-second Street in Manhattan to Hunters Point in Queens, into a subway that would proceed to Astoria in northwest Queens and to Woodside in the center of the borough. Queens had no rapid transit service, and this new IRT line promised to open up enormous tracts of rural land for residential and commercial development. The BRT's planning acquired a broader, more citywide focus, too. Following Edwin S. Winters' retirement as BRT president in mid-February, the new head, T. S. Williams, emphasized building a subway with a metropolitan orientation rather than one that would be merely a distribution system for Manhattan-bound passengers. The most significant product of Williams' thinking was a design for a trunk line that would run up the spine of Manhattan from the Battery to Central Park South.[57]

On June 13, 1911, the transit committee released a document, the McAneny report, that outlined a structure for the new subways, which became known as the dual system. The McAneny report formally proposed that the municipality join with the IRT and the BRT in constructing eighty-seven route miles of subways at a total cost of $249.4 million. The city would furnish $123 million for construction of the dual system, while the IRT and the BRT would contribute $75.8 million and $50.4 million, respectively, for construction and equipment.

The McAneny report envisioned the creation of a new relationship between the public and private spheres which would guarantee that neither the IRT nor the BRT would exercise as much power as the Interborough had since 1904. The report proposed that the city own the dual contract subways and lease them to the companies for forty-nine years. To prevent a corporate power grab, the contracts would contain the recapture provision of the indeterminate franchise formula that enabled the municipality to take over a line ten years after operation began.

This report also advanced an elaborate formula for dividing the subways' gross revenues: First, the company would recover its interest and sinking fund expenses, then the municipality would receive a payment for its fixed costs, and then the city and the company would divide the remainder equally. There was one important difference between the BRT's and the IRT's profit-sharing plans. Because the Brooklyn Rapid Transit Company's old, privately owned elevateds were to be joined with its new, publicly owned subways as an integrated system (unlike the IRT's subways and elevateds, which were separate divisions), BRT President T. S. Williams agreed to treat the subways and the els as a single financial unit by "pooling" their receipts together. In return for consenting to this important arrangement, which gave the city the benefit of a gigantic integrated system for the price of several subways, the McAneny report recommended that the BRT be given an annual sum equal to the net profit of the company the year before its new system began operating. This sum, called a preferential, would be deducted from the subways' gross revenues before the BRT or the city subtracted their fixed charges. The preferential was intended to reduce the BRT's risk in building the new lines by guaranteeing that its profit from the enlarged network would be no less than its current profit. The preferential was also designed to reward the BRT for building so many railways in the outlying sections. Although this decentralized route structure would accomplish the progressives' goal of dispersing residents from the city core, it would generate less passenger revenue for the company than a more concentrated network. The preferential would thus compensate the BRT for acceding to this business inefficiency in the name of the public good. The preferential constituted a pragmatic, highly creative solution to the problem of balancing the conflicting interests of the business world and the public sector.

The McAneny report provided for no such special contingencies for the IRT, mainly because Belmont's company was making so much money from Contract No. 1 that no guarantee seemed to be needed. In fact, the committee had hoped to pool the receipts of the IRT's original subway with those of its new lines in order to subsidize the dual system routes that went through undeveloped territory. But the Interborough was making a big profit in 1911, and August Belmont did not want to dilute its income. In exchange for pooling IRT revenues, Belmont wanted the PSC to grant the Interborough the same preferential. The PSC rejected this demand, reasoning that the IRT's Contract No. 1 and No. 2 subways were already publicly owned and would be integrated with the dual contract lines in any case, and a guarantee of the IRT's earnings would have entailed a financial commitment almost double the size of the BRT's preferential.[58]

The Board of Estimate adopted the McAneny report on June 21, 1911, and the Brooklyn Rapid Transit Company accepted it six days later. Board of Estimate and BRT officials soon began to discuss contract language. George McAneny had made substantial progress. As *American* magazine observed, "McAneny devised a plan which will give the whole city the greatest possible transit relief in the least possible time," and yet it would also provide "a broad basis for the future development of the city in all its parts."[59]

But the Interborough rejected the McAneny report. August Belmont and T. P. Shonts wanted to preserve their subway monopoly and strongly resisted the BRT's participation. Belmont was also unhappy that his company had not been awarded a preferential by the transit committee; if the BRT's profits were going to be ensured, then so should the IRT's. To achieve his objectives, Belmont decided to play a waiting game. He thought the dual contract negotiations would eventually collapse and the Interborough would then be able to impose its own terms. In July 1911, Belmont told Lord Rothschild:

> Summing up [our cables], it means that really nothing will come out of either the Brooklyn Rapid Transit or the Interborough negotiations, so far as the two corporations are concerned, in the immediate future. The Interborough is absolutely out of it already, and the agreement which the Board of Estimate has decided to enter into with the Brooklyn Rapid Transit Co. will be resisted both by the mayor and taxpayers, so that litigation will check and ultimately destroy it. . . . I, therefore, look to an effort, at some future time, to re-enlist the help of the Interborough on the subject. We will not only have to wait, but it is part of wisdom and dignity to do so. I doubt if the complete collapse of the present plans can be reached before the end of the year, or in any event the late Autumn. Under the circumstances, there is nothing for us to do but to operate our property and lay by our surplus.[60]

But the Interborough and the Board of Estimate resumed negotiations in November 1911. In part, this change reflected Belmont's realization that the success of the Board of Estimate–Brooklyn Rapid Transit conferences spelled the end of the IRT's subway monopoly whether he liked it or not. More important, Belmont and Shonts used their monopoly power to win a critical financial concession from McAneny and the other city negotiators as the price of the IRT's acceptance of the dual contracts. In January 1912, the municipality granted the IRT a preferential amounting to 8.76 percent of its total investment in the old and new subways; in 1912 this sum equaled approximately $6.3 million for the Contract No. 1 and No. 2 lines, plus 6 percent of the capital to be invested in the dual system. This was a very good deal for the Interborough, which would now receive a huge profit before the city recovered anything for its fixed costs from the pooled revenues.

Some reformers, such as President John Purroy Mitchel of the Board of Aldermen and publisher William Randolph Hearst, strenuously opposed these terms of contract and blamed George McAneny for letting the Interborough rob the city once again. In their view McAneny's decision to award a preferential to the Interborough was nothing less than a sellout. Although these critics urged the Board of Estimate to reject the entire dual system deal, McAneny contended that this settlement was justified because the dual system would dramatically increase subway mileage and expand the scope of built-up area. McAneny carried the day, and the Board of Estimate and Apportionment approved the dual system on May 24 with only Mitchel dissenting.[61]

Although court challenges and the passage of enabling legislation would consume another year, the McAneny compromise eliminated the last major obstacle to the dual contracts. On March 19, 1913, officials of the City of New York and the IRT and BRT signed Contracts No. 3 and No. 4. The two companies agreed to lease their city-owned subways for forty-nine years and pay for part of the construction charges and for the entire cost of equipment ($60 million for the BRT and $77 million for the IRT, as compared to the municipality's share of $152 million). The contracts contained the recapture, pooling, and preferential provisions.[62]

The dual system was a stupendous enterprise that dwarfed New York's existing transit system. (See Map 3, The Dual System, 1913.) It would double the length of New York's rapid transit network,

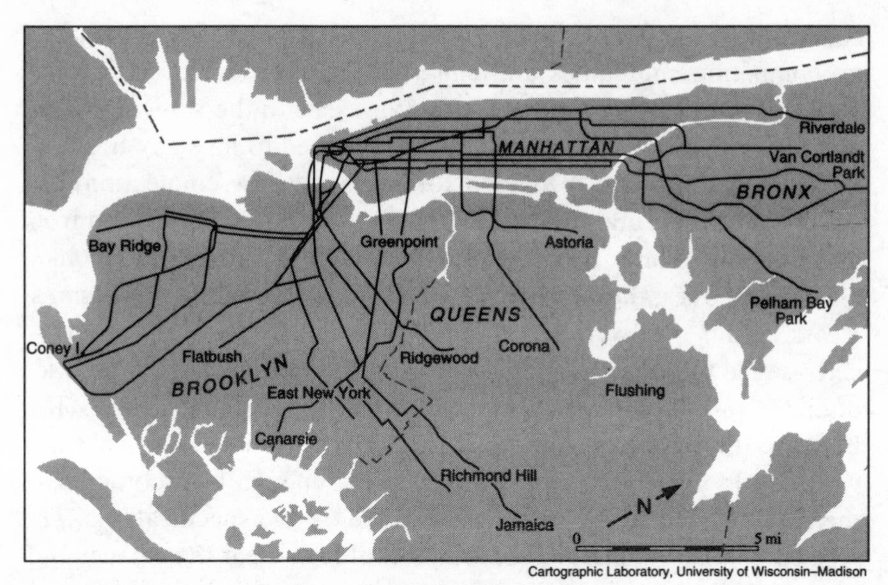

3. The Dual System, 1913.

increasing its single trackage from 296 to 619 miles. It would more than double the number of trains that the rapid transit system could provide in one direction per hour, from 352 to 851. And it would make New York's rapid transit system the world's leader.[63] By 1920, New York would surpass London and rank as the largest rapid transit system in the world. New York boasted 201.8 route miles of subways and elevateds in 1920, more than London (156.6 miles), Chicago (70.9), Paris (59.4), Boston (22.2), Berlin (21.7), and Hamburg (20.2).[64]

The size of this network was awe-inspiring. If the total track mileage of this system was laid end to end, the railway would stretch from New York City to Knoxville, Tennessee. If a passenger boarded a train at 9:00 one morning and traveled over the entire system at normal speeds, he would not return to his starting point until twenty-one hours later, at 6:00 the following morning. If the subways and els were used to full capacity for a twenty-four-hour period, they could have accommodated 35 million people, one-third of the nation's entire population.[65]

The dual system was not unflawed. Perhaps its greatest shortcoming was that it was divided into two separate networks managed by competing companies and charging separate fares. This rivalry between the IRT and the BRT was expressed in the subways' engineering design. Since neither company wanted to lose any traffic to its rival, the planners provided for relatively few connections between their lines, and the handful of passageways that did lead from one company's lines to the other's were long, narrow, and inconvenient. Moreover, the IRT and BRT tunnels had different dimensions. Years earlier the BRT had adopted a design for an elevated car—sixty-seven feet long, ten feet wide—corresponding to the standards of steam railroads; these specifications were larger than the IRT's. Because the BRT's dual contract lines would combine new subways with its old els, the tunnels had to be big enough to accommodate the el trains, and so the company retained the old specifications. The result was that the larger BRT trains could not clear the smaller IRT tunnels. If the two subway systems were ever unified, this mismatch would prevent integration with the old lines and prohibit joint orders for rolling stock.[66]

But the dual system's advantages far outweighed its disadvantages. Out of the mountains of red tape and from the piles of endless proposals and conference reports, a huge expansion scheme that fundamentally changed the life of the city finally emerged. This success was due primarily to George McAneny. Using his superb business and political connections and his unrivaled knowledge of New York City, McAneny tempered the progressives' moralistic opposition to cooperation with the subway companies, took advantage of the regulators' technical expertise, and secured the use of the IRT's and BRT's capital for the new lines. George McAneny made the political system function about as well as it possibly could, benefiting the entire city.

The dual system completed the enormous task of unifying Greater New York that was begun by the Contract No. 1 and No. 2 subways—leaping across the East River, binding Queens and Brooklyn to Manhattan and the Bronx, and opening thousands of acres of new land for development. "The time has come," a journalist said in *Munsey's* magazine, "when old Father Knickerbocker, sore from generations of jostling and car-crowding and with his shins barked from 'stepping lively,' can behold a vision of rapid transit deliverance.''[67]

By the time most of the dual system was finished in the early 1920s, a New Yorker could spend the morning swimming or riding the roller coaster at Coney Island, travel to Yankee Stadium for a baseball game that afternoon, and then head to Times Square for a movie that evening. Three fabulous forms of entertainment, located miles apart in three separate boroughs, were tied together by the most magnificent subway in the world.[68] Before the dual system subway, such cross-borough travel would have been extremely difficult and time-consuming; after it opened, New Yorkers grew so accustomed to traveling from one end of the city to the other that they forgot how hard it had once been.

7

ACROSS THE EAST RIVER

The dual subway system succeeded in overcoming the river barriers that separated Manhattan from the Bronx, Queens, and Brooklyn. When the dual contracts were signed in March 1913, the existing underground railways included only two crossings of the Harlem River to the Bronx and one of the East River to Brooklyn. By contrast, the dual contracts provided for eight crossings of these rivers. Of the eight lines, seven would span the East River, a half-mile-wide tidal strait that had impeded travelers for nearly three centuries. The new subways would breach Manhattan's river barriers and stimulate development in the far reaches of the city.[1]

Actually, one of the new East River tunnels antedated the dual contracts by three decades. This was the Steinway tunnel, which ran from Forty-second Street in Manhattan to Hunter's Point in the Long Island City section of Queens. The Steinway tunnel opened in 1915 as a link in the dual system's Queensboro route, but it had been conceived in the 1880s as part of a grandiose scheme to create a transriver freight and passenger railroad. The long, convoluted history of the Steinway tunnel reveals some of the engineering, financial, and political problems that thwarted New York's early underwater transit projects.[2]

The Steinway Tunnel

On July 30, 1887, a group of New York City capitalists incorporated the New York and Long Island Railroad Company (NY & LIRR). Led by financier Malcolm W. Niven, these investors wanted to build a railroad across the East River in order to solve a grave regional transportation problem. Spread across several big islands (Manhattan, Long, and Staten) and parts of the mainland in New Jersey and what is now the Bronx, the New York region was split by such major waterways as the Hudson River, the East River, and New York Harbor, all of which disrupted the movement of people and goods. Of all the railroads that served this booming metropolis in 1887, only one, the New York Central, entered Manhattan; all the others terminated on the New Jersey side of the Hudson River or the Long Island side of the East River. There were no bridges or tunnels across the Hudson between New Jersey and Manhattan; indeed, the closest bridge over the Hudson River was located seventy-five miles upstream in Poughkeepsie. The Brooklyn Bridge, completed in 1883, was then the only direct connection across the East River. The absence of such bridges and tunnels prolonged passenger journeys and raised transport costs.[3]

The New York and Long Island Railroad intended to correct this situation. It was conceived as a terminal railroad that would carry freight and passenger traffic across the East River, mainly between New York Central's depots in Manhattan and the Long Island Railroad's terminal in Queens County. Starting at Eleventh Avenue and Forty-second Street in Manhattan, near the New York Central's freight line (on Eleventh Avenue) and close to the Hudson River docks, the NY & LIRR's route would continue across Forty-second Street and past New York Central's main passenger station on Park Avenue. The route would continue across Forty-second Street and cross the East River to Long Island City, where it would connect with the Long Island Railroad's terminal in Hunter's Point. The NY & LIRR's route (including several branch lines) would be 5.6 miles long and cost approximately $11.7 million to build.[4]

Piano manufacturer William Steinway closely followed New

York City's and Long Island's activities from the outset. Before he played a major role in the development of New York's first subway, Steinway was best known as the president of the famous Steinway & Sons piano company and also as a booster of Queens County. After Steinway & Sons built a sawmill, a foundry, and a pianoforte factory there in the early 1870s, he promoted real estate and operated a street railway in Queens. Convinced that the region had great potential for growth, William Steinway responded to the second call for capital in July 1891 and soon became the railroad's largest stockholder. The tunnel that was to be drilled below the East River would thus carry his name.[5]

By 1891 the City of New York and Long Island City (which were separate municipalities before the consolidation of the five boroughs in 1898) had both approved the NY & LIRR's plans. The State of New York had also granted the NY & LIRR permission to tunnel beneath the East River. The NY & LIRR franchise, like so many others issued during the late nineteenth century, gave the company perpetual use of its right of way through the two cities and under the river. Similarly, the franchise imposed minimal fees on the company in exchange for these potentially valuable privileges. The City of New York required that the NY & LIRR remit 3 percent of the gross revenues it earned within the city limits, the State of New York levied a one-time fee of $500, and Long Island City did not charge anything.[6]

On June 7, 1892, laborers for the Inter-Island Construction Company began digging the tunnel. They started by sinking a vertical shaft in Long Island City, at the junction of Vernon, Jackson, and Fiftieth avenues, two blocks from the East River. In December 1892 the men reached the bottom of the shaft, at a depth of one hundred feet, and started drilling horizontally toward the river. Although the crews were making steady progress, the project was jeopardized by an accident that occurred on December 28, 1892. At around 7:00 that morning, foreman Peter McEntee and several workers carried eighty-seven one-pound dynamite cartridges from a storage shanty on the meadows east of Van Alst Street to a boiler room located near the opening of the tunnel shaft. McEntee and his men put the dynamite in a special steam-powered heating box used to warm the dynamite cartridges. This was standard practice in cold weather: Dynamite froze at around forty degrees Fahrenheit and had to be thawed before it could be detonated. This time, however, something went wrong. Shortly before 8:00 A.M. the dynamite cartridges blew up. Because the explosion took place on the surface rather than below ground, it did not damage the tunnel or hurt anyone who was working in it, but it caused extensive damage on the street level. It knocked pedestrians off their feet, sent broken window glass flying through the street, and shook building foundations for blocks around. It wrecked several residential and commercial buildings and set fire to a row of flats on Jackson Avenue. Five people died and dozens were injured.[7]

The accident ruined the New York and Long Island Railroad. Overwhelmed by legal claims demanding compensation for personal injuries and property damage, the NY & LIRR could not raise the funds needed to resume construction. Investors who were already wary of financing such a risky project now became further alarmed. The December 1892 explosion was indicative of the inadequacies of subaqueous tunneling in the late nineteenth century, a general engineering problem that was not confined to the Steinway tunnel. Twelve years earlier DeWitt Clinton Haskin had attempted to bore a railroad tunnel under the Hudson River from New Jersey to New York City, only to be stopped by an explosion that claimed twenty lives. The NY & LIRR and Haskin's Hudson River Tunnel Company were both casualties of the primitive engineering of late nineteenth-century tunnel building.[8]

For ten years the excavation site lay dormant. Then early in 1902 the Interborough Rapid Transit Company bought the NY & LIRR. This acquisition by IRT president August Belmont was part of a larger plan to dominate mass transit in the borough of Queens. Abandoning the NY & LIRR's scheme of creating a transriver terminal railroad stretching from Manhattan to Queens, Belmont intended to concentrate on the tunnel itself, completing it and equipping it for electric streetcars. The following year he purchased the electric New York & Queens County Railway, a forty-one-mile network. When the Steinway tunnel was finished, August Belmont would connect it with the New York & Queens County. Trolleys would then be able to start at Grand Central (close to the IRT subway station) and shoot through the East River tubes to Astoria, Corona, Flushing, Newtown, College Point, and Whitestone.[9]

In July 1905 the Degnon Construction Company began excavating a shaft at Fourth and Front streets in Long Island City.[10] Tunnel building proceeded more smoothly than it had in 1892, for underwater engineering had made great advances during that interval. Two English engineers, Sir Benjamin Baker and James Henry Greathead, had designed an improved tunneling shield that enabled headings to be driven longer, more quickly, and more safely than before. One sign of this technological progress was the large number of tunnels dug in New York City after the turn of the century. Between 1900 and 1910 the Interborough Rapid Transit Company finished one tunnel to the Bronx and another to Brooklyn, President William G. McAdoo of the Hudson & Manhattan Railroad Company revived DeWitt Clinton Haskin's failed enterprise and built two separate tunnels to New Jersey, and the Pennsylvania Railroad bored tubes below the Hudson and East rivers. Marveling that sixteen tunnels had been completed or were being built across the Hudson, Harlem, and East rivers in 1906, journalist Arthur B. Reeve remarked that "Manhattan may be described as a body of land surrounded by tunnels."[11]

The Steinway tunnel would consist of two tubes fifteen and a half feet wide and about six thousand feet long. Approximately twelve hundred workers worked on it. Most of the foremen appear to have been Irish, while the laborers included Italians, Poles, African-Americans, and others.[12] The tunnel was driven from four places. The first shaft (where work had already begun) was located near Fourth and Front streets in Long Island City; it took the place of the New York and Long Island's old shaft, which had been abandoned. The second shaft was at the foot of Forty-second Street in Manhattan, while the third would be at Third Avenue and Forty-second Street. The fourth shaft would be sunk through Man o' War reef, just south of Blackwells Island (now Roosevelt Island) in the middle of the East River.[13]

The shaft on Man o' War reef was particularly treacherous. Because the East River surrounded this site, it was constantly in danger of being flooded. To keep this from happening, five wood-lined, watertight cribs were installed in the shaft and equipped with air compressors; these compressors raised the atmospheric pressure high enough so that water could not seep in. Compressed air did not have to be used until much later in the Long Island City and Forty-second Street headings; both went through solid gneiss rock for so long that the air pressure did not have to be raised until far out in the river channel.[14]

The sandhogs worked under difficult conditions. Just getting to the job was difficult. To reach the rock face from the Long Island side, they walked down four flights of stairs to the bottom of the shaft and started trudging downhill toward the river. The narrow, muddy passageway was strewn with tools and filled with compressed air conduits, water pipes, and telephone and electric lines. Far out under the East River they entered at the air lock, an iron chamber that divided the zone of regular atmospheric pressure from that of compressed air. When the air pressure inside the lock equaled the pressure at the end of the tunnel, the men resumed their journey. Beyond the lock, the air was stuffy, moist, and warm. Finally, the workers came to the face where, under the protection of the shield, gangs drilled holes for the dynamite, triggered the charges, cleared the mud, sand, and rock that had been dislodged by the blast, and assembled the tunnel's wall.[15]

Tunneling was dangerous. Even though the East River was narrower than the Hudson River, it posed greater engineering problems, mainly because its bottom contained a greater variety of soil strata than that of the Hudson River, which was mostly silt. The most common material in the East River's bed was solid gneiss rock, which was perfect for tunneling, but in some places the gneiss was badly fissured and decomposed, with many pockets of mud, sand, and gravel. In addition, there was a ridge of soft dolomite limestone in the west channel. This geological diversity was dangerous because accidents often occurred during the transition from one type of strata to another. The Steinway tunnel thus gained a reputation for an unusually high number of blowouts and floods. One such accident took place below the river's western channel on July 2, 1906. The thirty sandhogs who were working in the tunnel had been cutting through solid gneiss when they encountered a shelf of sand and broken rock. One man described what happened next:

> Sand or ground-up stone began pouring in at about 10 o'clock [A.M.], and the tunnel lights grew dimmer. Then came a little stream of water, and it kept getting bigger, until finally a big heap of sand and rock came in and we knew that we had hit a shelf of sand. . . . Then came a hissing sound, and we knew that the water was coming at us. One of the men shouted "Danger!"[16]

The workers were trapped at the face of the rock, over 100 feet below the river and 230 feet from the mouth of the Manhattan shaft. As the water rushed in, moreover, the electric lights failed. So in near total darkness they scrambled back up the passageway, trying to save their lives. All survived; twenty-eight reached the air lock, and two made it to an emergency lock. The tunnel was flooded back to the air lock, and the project was delayed for three weeks until the water could be pumped out.[17]

But despite such floods, the tunnel progressed. On May 16, 1907, the workers who were drilling the two ends of the tunnel's north tube met in the middle of the river and connected their headings. Three months later, on August 8, the headings of the south tube were joined. Tracks were then laid and the overhead trolley wires strung.

On September 24, 1907, streetcar number 601 made the first official trip through the tunnel, carrying August Belmont and T. P. Shonts of the Interborough, Michael J. Degnon of Degnon Construction Company, several Public Service commissioners, and other dignitaries. The streetcar went from the Jackson Avenue station in Long Island City to Grand Central in less than four minutes, a dramatic improvement over the thirty-five or forty minutes needed to make the same journey via ferry and trolley. This extraordinary reduction of traveling time presaged major changes for Queens.[18]

Trains Meadow

Queens was primarily rural in 1900. It had 152,999 inhabitants that year and accounted for only 4 percent of the city's total population.

Although Queens was the largest of the five boroughs in terms of physical area and contained 35 percent of its total land mass, it was sparsely settled. The population density of Queens was 2 people per acre in 1900 compared to 16.8 for the entire city. Despite the presence of manufacturing centers, such as Long Island City, and market towns, such as Jamaica and Flushing, most of the borough's landscape consisted of fields, meadows, and woods. Bucolic Queens was so far removed from the hectic pace of city life that contemporary guidebooks encouraged Manhattanites to take day trips there, much as later guidebooks would recommend excursions to Long Island's Suffolk County, upstate New York's Catskill Mountains, and Connecticut's Litchfield County.[19]

One person who had firsthand knowledge of rural Queens was John H. Hendrickson. An avid hunter and outdoorsman, Hendrickson traveled to every corner of the borough in search of game birds —the hills and fields off Long Island City's Thomson Avenue for field plover; the cornfields near Calvary Cemetery for bobolinks; the freshwater ponds in Newtown and Flushing for snipe; and Long Island Sound's saltmarshes for ducks.

One of Hendrickson's favorite hunting grounds was Trains Meadow. Located between the settlements of Woodside and Corona in north-central Queens, Trains Meadow had escaped urban development and remained a rolling landscape of fields, meadows, and streams in 1909. The entire five-hundred-acre tract contained no more than two dozen or so houses. There were two big farms—the Barclay and Leverich places—and four or five truck farms that grew corn, peas, beans, and other vegetables for city markets in Manhattan and Brooklyn. Barns, dairies, cow sheds, corn cribs, and beehives dotted the fields.[20] Trains Meadow was extremely well watered and had many marshes and brooks; Sackhickniyeh brook, the largest, was six to eight feet wide and three or four feet deep in places. According to one account, the name Trains Meadow was a corruption of an older phrase, "Long Drains Meadow," derived from channels that had been dug to drain the wetlands.[21]

Hendrickson sometimes went hunting in Trains Meadow two or three times a week. He woke up early on Saturday or Sunday morning and put on his L. L. Bean outdoor clothing. Carrying his shotgun and accompanied by his dog Roger, he left his house at 130 Twelfth

Street in Long Island City, walked over to Jackson Avenue, and waited for a New York & Queens County trolley. Hendrickson could make it from Long Island City to Trains Meadow in a half hour or forty minutes; the New York & Queens County's red cars ran straight out Jackson Avenue, skirting the meadow's southern edge. After stepping off the trolley, Hendrickson immediately plunged into the brush and began looking for prey. In the spring and sometimes in the fall he hunted snipe, a brown-colored wading bird that fed on the banks of Sackhickniyeh brook and in the marshes. In the fall he shot woodcock in the wetlands' bushes and weeds.[22] Hendrickson relished his trips to Trains Meadow, praising it as "one of the most picturesque and beautiful" places in Queens; he enjoyed tramping through its "groves of trees and well-cultivated fields [and] small patch[es] of meadows."[23]

For all his appreciation of Trains Meadow's natural beauty, Hendrickson was hardly a country bumpkin. To the contrary, he was accustomed to an urban way of life. A well-educated, cultivated man who enjoyed the sociability of the town, Hendrickson lived and worked in Long Island City, the largest and most heterogeneous community in Queens. To get to the meadows he relied on the electric streetcar, a relatively recent product of modern technology. And Hendrickson made use of his hunting trips to participate in one of the country's most dynamic industries—the popular press. After returning from the countryside, he sat down at his typewriter and wrote a rough account of the day's events; later on he refined these diarylike entries into articles and submitted them to *Forest and Stream* magazine for publication. *Forest and Stream* was one of a number of periodicals that had recently sprung up in response to a growing public interest in the out-of-doors. As a regular contributor who was always looking for material for a piece, Hendrickson experienced nature in a disciplined, highly self-conscious manner that was characteristic of city life at its most sophisticated. Hendrickson's use of Trains Meadow was part of a phenomenon that historian Peter Schmitt identified as the "back-to-nature movement." Like many other Americans, Hendrickson valued the countryside for providing him with the authenticity missing from his everyday routine and for renewing him spiritually. By escaping to Trains Meadow on the weekend or after a day in his law office, Hendrickson returned home fortified, ready to face the city's challenges.[24]

Hendrickson romanticized rural Queens. He regarded "rural" and "natural" as synonyms and perceived the borough's farmland as a pristine land. In his diary entries and magazine pieces, for instance, he tended to concentrate on the area's marshes and brooks, while overlooking the farms and roadways that attested to the area's ties to the urban economy. In reality, Trains Meadow was already on the outer edge of the metropolis and was steadily being drawn closer to the urban orbit: The same electric railway that whisked Hendrickson to the meadows carried other people there, too. As early as 1900, he noticed telltale signs of development: The hard dirt surface of Jackson Avenue was paved with macadam; a quaint plank bridge, which had an appealingly rickety guardrail and crossed Sackhickniyeh brook, was replaced with a nondescript stone culvert; and the Citizens' Water Supply Company installed a pumping station to pipe water to Long Island City and dry up the wetlands. To Hendrickson these public works improvements threatened Trains Meadows' pastoral beauty. "As I advance I notice on every hand the sad changes that time and man—principally man—has wrought," he mourned in 1900. "A few years ago this country was one of the most picturesque I had ever visited, but now everything is changed."[25]

But this was only the beginning. By 1909, Hendrickson recognized that Queens was on the verge of a sweeping urban transformation that would lead to the "rapid colonization of our whole island" and obliterate the rural landscape. "No spot seems to [*sic*] barren, no place too remote to erect a dwelling," he lamented, "and our beaches and meadows, formerly the home of the plover or meadow hen, are now being occupied for residential purposes at a rate that dazes the lover of nature." He wondered "where our baybirds and waterfowl will then find shelter."[26]

John H. Hendrickson had good reason to be worried. In making Queens a borough of the second largest city in the world, the consolidation of greater New York in 1898 established a political framework for comprehensive urbanization. The borough began to grow rapidly following the completion of three major transportation projects that linked Queens directly to Manhattan. One was Pennsylvania Railroad's station, tunnel, and bridge complex. Built from 1905 to 1917, this massive project included a tunnel from Manhattan to Queens, the Sunnyside Yards in Long Island City, and the Hell Gate Bridge between Queens and the Bronx. The second was the Queensboro Bridge from Fifty-ninth Street in Manhattan to Bridge Plaza in Long Island City, opened in 1909.

The third project was the Steinway tunnel. Even though the tunnel had been completed shortly before streetcar number 601 made its ceremonial run through the tubes in September 1907, its formal opening was delayed by a dispute between the government regulators and the Interborough Rapid Transit Company over the validity of the IRT's franchise. The Interborough had built the tunnel under the terms of the New York and Long Island's old franchise. Issued during the laissez-faire period of the late nineteenth century, this franchise granted the private corporation perpetual control of a public right-of-way, allowed it near complete freedom from government supervision, and imposed modest fees on it. Such generous grants of power to a private business were not uncommon during the 1890s, but the construction of the Steinway tunnel took so long that the political situation changed by the time it was completed. The Steinway tunnel thus became entangled in the general conflict about government regulation of business that was a major concern of the progressive era. It also became caught in the Interborough Rapid Transit Company's battles with the Public Service Commission over its monopoly of rapid transit in Manhattan and its refusal to build new subways; until these critical issues were settled, the PSC refused to sanction the Steinway tunnel's opening. Only when the IRT and the PSC resolved their larger differences, with the signing of the dual contracts in March 1913, was the Steinway tunnel released from limbo. Under the terms of Contract No. 3, the IRT agreed to sell the Steinway tunnel to the City of New York for $3 million and operate it as part of its Queensboro subway route.[27]

Originally equipped for electric streetcars, the tunnel was then reconstructed to handle the IRT's subway trains. The tunnel finally opened on June 22, 1915, twenty-three years after construction had started and eight years after trolley number 601 had gone through it.[28] The extension of this first subway line to Queens, along with the completion of the Pennsylvania Railroad's improvements and the Queensboro Bridge, transformed many parts of the borough, including Trains Meadow. The man who developed Trains Meadow was Edward A. McDougall.

Jackson Heights

Born in Brooklyn on July 16, 1874, Edward A. McDougall entered real estate after leaving public school. In 1906 he formed a partnership that built housing in outlying parts of Queens, including Kissena Park in southern Flushing, Terminal Heights in Woodside, and Elmhurst Square in central Queens.[29]

On August 12, 1909, McDougall incorporated a new company, Queensboro Corporation, to develop Trains Meadow. Later that year McDougall started negotiating with Trains Meadow's landowners in an effort to buy their property. Queensboro took title to its first parcel, the 128-acre Barclay-Dugro tract, in January 1910. By 1914, Queensboro had acquired roughly 350 acres there.[30]

McDougall gave Trains Meadow a new name, Jackson Heights. He took the "Jackson" from Jackson Avenue, the area's chief east-west artery prior to the extension of Northern Boulevard. McDougall's adoption of the word "Heights" was more imaginative. Rather than describing a striking geographical feature, McDougall was probably drawing upon a well-known association between social exclusivity and physical elevation that had been established by such fashionable places as New York's Brooklyn Heights, Boston's Beacon Hill, and San Francisco's Nob Hill.[31]

Jackson Heights grew slowly before World War I.[32] The handful of buildings were surrounded by vacant lots that presented a raw, unsightly appearance. Most of the streets were dirt roads that led nowhere; only a few had been paved. "The first residents of Jackson Heights were true pioneers," architectural historian Daniel Karatzas concluded.[33]

Jackson Heights' main drawback was its inaccessibility. To reach mid-Manhattan, for instance, residents had to catch a streetcar to Hunter's Point and then take a ferry across the river to Thirty-fourth Street. Because this journey took an hour or an hour and a half, most commuters would not consider moving to Jackson Heights. But as McDougall realized, the Interborough's Queensboro subway promised to bring Jackson Heights within easy traveling distance of Manhattan. Ever since the dual contracts were signed in 1913, McDougall insisted that the Queensboro line would guarantee Jackson Heights' success. His confidence was justified. Once the line to Jackson Heights opened on April 21, 1917, passengers could board an IRT train at the Eighty-second Street station—the twelfth stop in Queens—and arrive at Grand Central Terminal in twenty or twenty-five minutes. Afterward, Queensboro's real estate ads regularly boasted that Jackson Heights was only a twenty-two-minute subway trip from Grand Central.[34]

Between 1917 and 1919, McDougall devised a plan for Jackson Heights as an exclusive garden suburb for upper-middle-class New Yorkers. McDougall's conception was influenced by British town planner Ebenezer Howard's idea of creating garden cities that would combine the sociability of the town with the healthfulness of the countryside. As Howard outlined his thinking in his famous 1898 book, *Garden Cities of To-Morrow,* garden cities were new, rationally planned communities that would eliminate the physical chaos and social misery of the nineteenth-century metropolis.[35] Yet McDougall's scheme also belonged to an older American tradition of business city planning, a tradition that had been responsible for the development of earlier suburbs such as Llewellyn Park, New Jersey; Lake Forest, Illinois; the Country Club District in Kansas City; and the Overbrook Farms section of Philadelphia. Unlike social reformer Ebenezer Howard, Edward A. McDougall was a profit-oriented businessman, and his Queensboro Corporation, a well-capitalized, vertically integrated company, was in the forefront of real estate practice. They represented a departure from the pattern of development that had followed the construction of the Contract No. 1 subway. In northern Manhattan and the Bronx, speculators such as Charles T. Barney had bought large tracts of undeveloped land before the prices rose and divided them into lots for sale to small-scale builders, who in turn erected tenements that were sold to investors. This city-building pattern was highly speculative and involved several different participants. By contrast, Queensboro Corporation conceived of Jackson Heights as a completely planned and largely self-contained development. To realize his vision, McDougall needed to control the entire city-building process. His Queensboro would buy the land, draft a master plan, subdivide the property into blocks and building lots, install streets, sidewalks, and utilities, build and market the housing, and establish a cohesive community.[36]

McDougall wanted to create a suburban haven within city limits. Since Jackson Heights was part of New York City, it could never be politically autonomous and self-governing. To McDougall, however, Jackson Heights nonetheless qualified as a suburb because its low population density, residential homogeneity, and distinctive way of life presented a striking contrast to the rest of the city, particularly Manhattan. Queensboro Corporation's advertising defined Jackson Heights in relation to Manhattan.[37] Queensboro claimed that Jackson Heights was located far enough from Manhattan "to miss its noise and congestion" and to preserve "the atmosphere of leisure and dignity which has fled Manhattan." Yet, as the brochures and ads were quick to note, Jackson Heights offered all of Manhattan's municipal services and shared its cultural sophistication. "It is quiet as a country town but decidedly metropolitan in its taste," Queensboro reassured home buyers who were nervous about being too far from the action.[38] Here, middle-class men could go to their Manhattan offices in the morning and return at night to an island of domestic peace and harmony where their wives fashioned the good life for them and their children.

Even though suburbs were usually associated with single-family, detached houses, McDougall decided to build mainly apartment houses. He promoted Jackson Heights as the nation's first garden apartment suburb. His concentration on apartments was due to New York City's costly land prices, high population densities, and tradition of multiple-unit dwellings. Under the terms of a cooperative ownership plan unveiled in 1919, residents could buy their own flats and become homeowners. This cooperative plan proved to be popular; by April 1921, six hundred families had bought apartments. Unlike the average New York City apartment dweller, many Jackson Heights residents owned rather than rented their flats.[39]

Queensboro Corporation achieved a high standard of apartment construction. It embellished the facades, stairways, and halls with ornate architectural details and provided parquet floors, fireplaces, sunrooms, built-in bathtubs with showers, and solid plaster construction in its flats. What distinguished Jackson Heights was that the block, not the individual building, became the main planning unit. Between 1920 and 1925, Queensboro Corporation erected eight apartment complexes that occupied an entire block apiece. Each block-long complex usually contained ten four- or five-story structures that had symmetrical heights and a uniform architectural style. The buildings were sited on the edges of the block, with five apartments lining one street and five lining the opposite street. This placement left the center of the block empty. Shielded from the street by the buildings and protected from traffic noise and pedestrians' stares, these interior spaces became semi-private oases of light, air, and greenery.[40]

McDougall enhanced Jackson Heights' appeal by providing residents with special amenities. Rather than developing every lot with housing, Queensboro reserved some of its land for community functions while retaining title to the property. For instance, Queensboro set aside plots for supervised playgrounds, a community garden, tennis courts, and a twelve-hole golf course (billed as "the nearest Golf Course to Herald Square in New York City").[41]

As part of its well-organized approach to real estate development, Queensboro Corporation sought to promote social cohesion and foster a distinctive way of life. McDougall conceived of Jackson Heights as a homogeneous suburb of like-minded middle-class residents who shared similar social backgrounds, occupational levels, and cultural values.[42] "Jackson Heights," Queensboro declared in 1919, was "a community of neighbors; an association of home-loving people whose aim in life is to promote fellowship, happiness and good will. Its watchword is cordiality and its creed is friendship."[43] The key to McDougall's efforts at community building was the cooperative ownership plan; he thought that home ownership would give people a permanent stake in Jackson Heights' future and foster middle-class values of order, stability, and harmony. In addition, the company established an intricate organizational structure intended to strengthen families and build a strong community. Queensboro, for instance, supported art, literary, and women's clubs, sponsored tennis, golf, bowling, and basketball leagues, and encouraged children's activities. It also supported a newspaper, the *Jackson Heights News,* which carried items about local residents and reinforced Jackson Heights' identity as a distinct place.[44]

As a strong private corporation that intervened in all aspects of Jackson Heights, Queensboro wielded great power over the lives of its inhabitants. It used its power to protect Jackson Heights' property values and reputation for suburban exclusivity. Queensboro thus enforced middle-class standards of decorum by prohibiting residents from shaking their mops or brooms out of their windows or from drying their wash on clotheslines.[45] Yet Queensboro's efforts at social control had a darker side that involved the selection of residents. In its brochures Queensboro claimed that its agents screened applicants "with a view to [ensuring] financial and social compatibility" with other residents and ensuring "common ideals and living standards."[46] Translated into plain English, these polite code phrases meant that Queensboro systematically excluded Jews from Jackson Heights. Its real estate advertisements proudly identified Jackson Heights as a "Restricted Garden Residential Section" and proclaimed "Social and Business References [Were] Required" to buy or rent there.[47] The exclusion of Jews was central to Queensboro Corporation's garden suburb ideal. Jackson Heights' solidly northern European character was part of a larger trend. At a time of growing national pressure to restrict immigration, especially from southern and eastern Europe, Queensboro's sophisticated promotion of Jackson Heights was designed to appeal to a Protestant sense of superiority over lesser breeds and to reinforce feelings of self-importance. During the 1920s and 1930s, anti-Semitic housing discrimination was not limited to Jackson Heights, of course. But although many other New York City landlords refused to sell or rent to Jews, the anti-Semitic restrictions gained special force in Jackson Heights as a result of Queensboro's tight hold on the area; due to the corporation's vertical integration, they were applied to an entire section of the city rather than a few structures. And these restrictions were effective; very few Jewish New Yorkers lived in Jackson Heights before the 1950s.[48]

Jackson Heights thrived during the 1920s. From 1923 to 1930

its population increased elevenfold, from 3,800 to 44,500. By 1929, Queensboro had built 135 apartment buildings and 500 single-family homes that had a total value of $50 million.[49] But the great depression devastated Queensboro. Three of its thirteen cooperative apartment complexes failed in the 1930s, while the prices of apartments, which had gone for $20,000 in the 1920s, sank to as low as $3,000—and still went unsold. With such little demand for housing, McDougall was compelled as early as 1929 to divide some of his seven-room flats into cheaper, less luxurious three- or four-room apartments. In an even more startling departure from his ideology of suburban exclusivity, he began offering furnished rooms for rent. Although the company began to recover in 1934, it remained in financial difficulty for the rest of the decade. As a result, Queensboro Corporation soon made a more fundamental change in its master plan, compromising its original vision of a garden suburb. In order to raise more money, McDougall decided to develop the land that had been kept open and parklike. Beginning in the late 1930s new apartments covered the golf course, tennis courts, community garden, and other open sites.

With the elimination of the parklike atmosphere, Jackson Heights no longer resembled either a garden or a suburb. It had become an urban neighborhood that was no longer insulated from the larger city.[50] Edward A. McDougall had conceived of Jackson Heights as a northern European suburban enclave within the most heterogeneous city in the world. Through skillful planning and shrewd marketing, Queensboro Corporation achieved this garden suburb ideal for two decades. Yet the corporation could not hold back larger urban forces forever; the neighborhood remained a political unit of New York City and continued to be connected to Manhattan via subway. In the end, when Queensboro was too weak to maintain its suburban ideal, New York City engulfed Jackson Heights.

Jackson Heights was only one of many communities that arose in the wake of the dual subway system. The new IRT and BRT subways stimulated the residential settlement of a vast crescent of undeveloped land that stretched from the north Bronx, across Queens, and through Brooklyn. This crescent encompassed neighborhoods from Westchester and Parkchester in the Bronx, to Elmhurst and Corona in Queens, and Canarsie and Bay Ridge in Brooklyn.

The emergence of these neighborhoods was part of an important demographic transformation that changed the entire city: the deconcentration of population from Manhattan to the outer boroughs. As Manhattan Borough President George McAneny and other progressive reformers had hoped, the dual system subways enabled New Yorkers to move from overcrowded Manhattan to new areas on the outskirts. Even though the total population of New York City increased 56 percent between 1910 and 1940, the number of people who lived in Manhattan fell 19 percent during this period, from 2,331,542 to 1,889,924. As the population of Manhattan began to drop in 1910, the outer boroughs reported dramatic gains. Between 1910 and 1940 the population of the Bronx shot up 309 percent, that of Queens jumped 218 percent, and that of Brooklyn increased 165 percent. The subways' importance as a catalyst of decentralization is revealed by the fact that 91 percent of all city residents lived within a half mile of a rapid transit railway in 1925. Clearly, people were moving out along the paths of the IRT and BRT lines.[51]

Although Manhattan remained the financial, governmental, and cultural capital of the city, the island was no longer its population center. In 1910, forty-nine of every one hundred New Yorkers lived on the twenty-three square miles of Manhattan Island; by 1940 only twenty-five of every one hundred resided there. In 1930, Brooklyn overtook Manhattan as the most populous of the five boroughs.[52]

For most Americans, and indeed for most New Yorkers who had forgotten how densely packed their narrow, splinter-shaped island had been during the nineteenth century, Manhattan remained the symbol of a crowded big city. And compared to other big U.S. cities such as Boston, Pittsburgh, Chicago, Atlanta, Phoenix, and Los Angeles, Manhattan certainly is thickly settled. But Manhattan is nonetheless far less crowded today than it used to be; from 1910 to 1940 its population density dropped from 161 people per acre to 130 per acre.[53] This transformation had a particularly great impact on such neighborhoods as the Lower East Side and Greenwich Village. These tenement districts, which had teemed with immigrants before

World War I, now began to empty as Jewish inhabitants followed the subways to Boro Park and Brownsville, and Italians took them to Bay Ridge and South Ozone Park.

The exodus of so many working-class residents to the outer boroughs altered the social makeup of subway passengers. Although all kinds of New Yorkers patronized the subway from the time that it opened in 1904, the ridership appears to have been weighted toward the upper middle class before World War I. The completion of the dual system changed that. Before World War I, many workers who had lived in Greenwich Village or on the Lower East Side walked to their jobs; in the 1920s, having moved to neighborhoods in the outer boroughs, they had to use the IRT or BRT to get to work. The result was that workers probably accounted for a larger proportion of total subway patronage in the 1920s than before.[54]

The shifting demography of subway ridership, together with a severe financial crisis that began during World War I, precipitated important changes in subway politics in the 1920s.

8

JOHN F. HYLAN AND THE IND

Crossroads

At the end of World War I, inflation forced a turning point in New York City rapid transit. It devastated the Interborough Rapid Transit Company and Brooklyn Rapid Transit Company, provoked intense political conflicts over attempts to bail out the two corporations, and changed the direction of subway planning.

Consumer prices advanced 56 percent between 1917 and 1920. In 1919, IRT president T. P. Shonts complained that in the previous three years the price of brake shoes had shot up 150 percent, a ton of steel had tripled from $30.00 to $90.00, and a ton of coal had gone from $3.23 to $6.07. Fuel costs alone added nearly $2 million to the IRT's operating budget. President T. S. Williams of the BRT claimed that the higher coal prices were taking $1 million more from his company per year. Wages and salaries had skyrocketed, too. In 1919, Interborough paid its employees $6 million more than in 1916.[1]

The sudden increase in costs created a crisis because the rapid transit systems were prohibited from raising fares. At the end of the nineteenth century the standard price on most American urban railways for a single ride of unlimited distance was a nickel. That, for instance, was the price of a trip on New York's elevateds in the

1890s. The straight nickel fare, as it was called, had been adopted for the subway almost automatically. Both the Rapid Transit Act of 1894 and the 1900 Contract No. 1 required the Interborough Rapid Transit Company to charge no more than five cents for a subway ticket. The nickel fare was retained in the dual contracts of 1913.[2]

Postwar inflation lowered the real value of the five-cent fare. In constant 1904 dollars, the value of a nickel plummeted to 2.6 cents by 1919 and then to 2.25 cents by 1920. While the fare lost 48 percent of its value from 1904 to 1919, the price of most other goods and services rose an average of 92 percent. Between 1904 and 1919 the cost of a man's necktie increased from fifty cents to $1.35, a pound of coffee from seven cents to twenty-two cents, a pair of girl's shoes from $2.00 to $5.50, and a copy of the *New York Times* from one penny to two cents.[3] For New Yorkers who were feeling pinched by surging food, clothing, and housing expenses at the end of World War I, the nickel fare was a welcome bargain.

But the depreciation of the fare was a disaster for the IRT and BRT.[4] The rapid transit systems were hit doubly hard by the nickel fare. They lost money because of inflation, of course, but they lost a second time due to the tremendous growth in mileage that had occurred since 1904. Because the Contract No. 2 extension and the dual system multiplied the total subways trackage sixfold, New Yorkers were able to travel much farther on a nickel ticket than before, increasing the length of the average ride. Since the nickel fare permitted trips of unlimited distance, the IRT and BRT received no extra revenue for these longer rides.[5]

By the end of the war the Interborough Rapid Transit Company and Brooklyn Rapid Transit Company were experiencing hard times. From 1917 to 1921 the Interborough's annual net income plunged more than 50 percent, to approximately $4.5 million. The Interborough almost became insolvent in 1918. Thanks to a surplus from its strong prewar performance, however, it survived this financial squeeze without having to enter receivership. But the Brooklyn Rapid Transit Company sustained huge operating deficits and was teetering on the edge of bankruptcy by Armistice Day 1918.[6]

This inflationary surge devastated railways throughout the United States.[7] Transit expert Delos F. Wilcox reported in 1919 that "the street railway business throughout the country is in a critical condition."[8] In a single two-month period in 1918, railways in 332 towns and cities discarded the straight nickel fare and raised ticket prices. By 1920 railways in 500 cities had hiked their fares. Many companies tried to save money by abandoning trackage, deferring routine maintenance, or delaying the purchase of new rolling stock, and some were forced to declare bankruptcy. In New York State, for instance, eleven railway companies, operating nearly one-quarter of the Empire State's total street railway mileage, were in receivership by 1920.[9] For the U.S. transit industry, World War I divided the earlier period of healthy profits from the later period of chronic deficits.

In addition to inflation, transit companies confronted a challenge from a potent new rival: the automobile. The earliest gasoline-powered vehicles had been produced in Europe during the 1880s. In September 1893, Charles E. and J. Frank Duryea, two brothers who worked as bicycle mechanics in Springfield, Massachusetts, built the first satisfactory American horseless carriage. With its large, spindly wheels and surrey top, the Duryea brothers' one-cylinder vehicle resembled a horse-drawn buggy and hardly seemed capable of revolutionizing transportation. For the rest of the 1890s motor vehicles remained little more than experimental curiosities or playthings of the wealthy. But after 1900 manufacturers broadened the automobile's appeal by making design and production improvements and by lowering car prices. The new era of mass consumption was symbolized by Henry Ford's famous Model T, introduced in 1908. Easy to operate, mechanically reliable, simple and economical to repair, the Model T has been called the first car of the multitudes. The dissemination of the automobile accelerated after World War I when Detroit's factories began churning out millions of cars and prices tumbled. The Model T sold for $393 in 1923, a bargain compared to its introductory price of $850 in 1908. Consequently, the number of motor vehicles registered in the United States more than doubled, from 9 million in 1920 to 20 million in 1925. By 1925 there was one motor vehicle for every five Americans.[10]

As private automobiles began competing with mass transit for passengers and street space, hundreds of cities and towns confronted important policy choices regarding traffic and parking regulations, highway construction, transit subsidies, fares, and other matters. Although most cities decided in favor of the auto, New York City was different. Because of its arduous physical geography and huge population, it had always been unusually dependent on mass transit —and rapid transit lines, not surface railways, carried the majority of riders. Although the subways and elevateds lost construction funds to highway projects beginning in the 1920s and then lost passengers to private automobiles after World War II, they did not compete directly with cars for street space, as happened in Chicago, for instance. And unlike trolleys, the subways were never in danger of being physically abandoned. Due to New York's combination of economic growth and daunting geography, plus its tremendous capital investment in underground mass transportation, the subways were permanent fixtures.[11]

The evolution of New York's subway thus began to diverge more sharply from that of other North American urban railways in the 1920s. As Los Angeles, Chicago, and other cities eliminated street railways during the interwar decades, New York City built a new rapid transit network, the Independent Subway System (IND) that became the third phase of subway construction undertaken in New York. Even though the IND was the last major subway completed in New York, rapid transit has remained unusually important. In 1989 the subway carried one of every nine mass transit passengers in the entire United States. In 1990, 46 percent of all New York workers used the subways to commute to their jobs, a figure far greater than that of any other U.S. city.[12]

The inflationary squeeze on operating profits and the competition of the automobile combined with a third force in New York City to undermine the city's commitment to the subway system and begin the process of decline: the intense politicization of the subways. Rapid transit had always been part of the political arena in New York City, but a decisive shift occurred after World War I when the forces of conflict overcame the forces of cooperation, and compromise became much more difficult. The postwar financial crisis increased the friction between the subway companies and the city and led to angry confrontations over the rate of fare, corporate reorganization, and the construction of new subways.[13]

The man who was primarily responsible for politicizing the subways was Mayor John F. Hylan, a fiery demagogue who dominated transit affairs during his two terms in city hall, from 1918 to 1925.

John F. Hylan

John F. Hylan was born on April 20, 1868, near the Catskill Mountains town of Hunter in upstate Greene County. His father had fled to the United States at age seven from his famine-ravaged home in County Cavan, Ireland, and later fought in the Civil War with the 120th New York Infantry Regiment; his mother was descended from a Welsh family that had arrived before the American Revolution. The Hylans farmed a sixty-acre tract of thin, worn-out soil. Unable to produce a big cash crop for market on this meager parcel, the Hylans could never get out of debt. For the family to survive, all five children had to contribute. As the oldest boy, young John worked from sunup to sundown milking cows, hoeing potatoes, mowing hay, cradling grain, and building stone walls. Despite his hard work, Hylan's parents could not afford to give him more than one pair of boots a year, and he often went barefoot. For the Hylan family, education was a luxury. At most, John F. Hylan was able to spend four or five hours a day in the local one-room schoolhouse for five months of every year; he never made it beyond the elementary level.

At fourteen Hylan started working for the Stoney Cove, Catskill & Kaaterskill Railroad to supplement the family income with his daily wages of $1.10. He carried drinking water to the gangs that laid track, and he repaired roadbeds damaged by spring floods. He also ran errands for train conductors. Hylan soon advanced to the better paying position of locomotive fireman, but he realized that his future would be bleak if he stayed at home. In 1887 nineteen-year-old Hylan decided to move to Brooklyn to make a fresh start. He left home with only $3.50, and on his first day in town landed a job as a track layer for the Brooklyn Union Elevated Railroad. Moving up to stoker and then to engine hostler, Hylan was promoted after two years to the elevated's top manual position, locomotive engineer. Earning the handsome pay of $3.50 a day, Hylan was able to get married, retire the $1,500 mortgage on his parents' farm, and support several siblings who had followed him to Brooklyn. At his wife's urging he enrolled in evening classes at Brooklyn's Long Island Business College to make up for his poor education. Then he entered New York Law School. For the next few years Hylan juggled his family duties, railway job, courses, and a clerkship in a Long Island City law firm. He received his LL.B. degree in October 1897.[14]

The month before his graduation from law school, Hylan had a painful experience at work one night: Driving his steam-powered elevated train at the Navy Street curve, he nearly hit a company superintendent named Barton. Writing about this incident in his autobiography thirty years later, Hylan denied all responsibility for it. Insisting that he had obeyed company rules by slowing his locomotive before it entered the curve, Hylan blamed the near-accident on Superintendent Barton, a "very old man" who had stepped out from behind a switch tower and started to walk across the tracks without looking both ways. "It was his own fault . . . and not mine," Hylan claimed. However, Hylan did concede that he had been "busily studying the law" for his upcoming bar examinations and that his mind might have wandered; and it is possible that his own carelessness may have contributed to the incident. In any event, a furious Superintendent Barton summoned Hylan to his office and summarily fired him. Hylan, who claimed that no other charge had been made against him in his nine years with Brooklyn Union Elevated, remained embittered by his dismissal for the rest of his life.[15]

Fortunately, Hylan had saved enough money to open a law office in the Bushwick section of Brooklyn. Business was slow at first—his practice consisted mainly of bill collections and minor criminal cases—and he earned only $24.00 the first month. Like many other struggling young lawyers, Hylan joined his local Democratic club, in Brooklyn's Twentieth Assembly District, in order to meet clients. He slowly rose in the party. In 1906, Mayor George B. McClellan named him a city magistrate, and eight years later Governor Martin H. Glynn appointed him to fill a vacancy in the Kings County Court. In 1916, Hylan announced his candidacy for a full seven-year term on the Kings County bench. Strongly endorsed by William Randolph Hearst's *Journal* and *American* newspapers, Hylan ran ahead of the entire Brooklyn Democratic ticket and received a plurality of thirty-seven thousand votes.

Hylan's big break came in the 1917 mayoral election. Newspaper publisher William Randolph Hearst was boosting Hylan for the mayoralty. Judge Hylan had attracted Hearst's attention four years earlier as a result of his strong opposition to the dual system contracts. Hylan had been very active in fighting these 1913 agreements, in part because he contended that the financial arrangements unduly favored the private subway companies, and in part because the dual system lines went mostly to undeveloped parts of the city. Hylan, representing a Bushwick constituency, wanted subways to go to the older, more built-up areas. Hearst admired Hylan's stand and became one of his supporters. Tammany Hall leader Charles F. Murphy reluctantly surrendered to pressure from Hearst and the Brooklyn Democratic organization and accepted Hylan as the party's nominee, despite Murphy's apprehension that Hylan was "too independent."[16]

The 1917 general election was unusually bitter and divisive. The incumbent, John Purroy Mitchel, had been elected on a fusion ticket in 1913 and was now running for reelection as an independent. As mayor, Mitchel had tried to reform the municipal administration by cutting spending and lowering taxes, a grimly efficient, penny-pinching approach that antagonized the poor immigrants who desperately needed steady jobs, decent housing, and cheap food. Mitchel's reelection campaign alienated still more New Yorkers. Vigorously supporting American intervention in World War I, Mitchel accused opponents of the war effort of toadying to the "Hohenzollerns and Hapsburgs" and being disloyal to the United States. John Purroy Mitchel's enthusiasm for the war infuriated New York's two largest voting blocs, the Irish and the Germans. By making patriotism the leading issue in a city where 66 percent of all residents were of foreign parentage, moreover, Mitchel inflamed the entire population.

John F. Hylan opposed the U.S. declaration of war against the Central Powers, a position that was popular with Irish and German New Yorkers. Yet Hylan's vote-getting appeal was imperiled by a third candidate, Socialist Morris Hillquit, who enjoyed a large following, especially on the Jewish Lower East Side. Hillquit denounced the war as a product of imperialism, called for the protection of civil liberties, demanded public ownership and operation of utility companies, and stumped for municipal distribution of food at cost. To cut into Hillquit's base of support, Hylan forcefully attacked "interests" such as John D. Rockefeller, New York Edison, Interborough Rapid Transit Company, and Brooklyn Rapid Transit Company. On November 6, 1917, Hylan scored a tremendous victory, taking 46 percent of the vote to Mitchel's 23 percent and Hillquit's 22 percent.[17]

With this win John F. Hylan completed a long and difficult journey from boyhood poverty to leadership of the nation's largest city. Tall and portly, with a large face, thinning red hair, a pencil moustache, and a stiff bearing, Hylan had an uncomplicated, naive view of the world. Red Mike, as he was nicknamed because of his hair color, clung tenaciously to his ideas, yet he often seemed hesitant and unsure of himself in social situations and before the public. In a profile of Hylan that appeared in *American Mercury* in 1924, writer William Bullock captured the mayor's awkwardness:

> A man of ambition, but apparently afraid or unable to stand alone; and so leaning on others, and so attempting to mask himself behind a stony reserve, and so hesitant to meet and fraternize with his fellows; slow and stumbling in his talk, and indecisive, on his own account, in action, and even in his walk; timid to the point of refusing a gesture when on the platform; coming before his audience like a man on stilts; standing there a bulky, seemingly inanimate figure, reading off in a sing-song monotone the words that others have written for him, and never venturing a word more. A queer hero, indeed.[18]

Master builder Robert Moses described Hylan as "a decent political hack" who "swelled instead of growing" in city hall, while Major General John F. O'Ryan, a prominent soldier and lawyer, called him "honest but befuddled."[19] Hylan sometimes did not appear to be in control of his own administration. Wits quipped that he gave his patronage to Charles F. Murphy and took his policies from William Randolph Hearst. He was, indeed, Hearst's mayor.[20]

In spite of his weaknesses, John F. Hylan helped transform New York subway politics. Elected chief executive at a critical time in the city's transit history, Hylan made mass transit his leading issue. As an urban populist who employed militant language, disdained the rights of property, and defended the "people" against the "plutocrats,'' the mayor spiritedly condemned the IRT and BRT as greedy, power-mad behemoths that double-crossed the public at every turn.

Hylan himself felt great personal hostility toward the subway companies. His animosity apparently stemmed from his bitterness about his ugly encounter with Superintendent Barton of Brooklyn Union Elevated in 1897. During his entire political career, Hylan expressed more rancor toward Brooklyn Rapid Transit Company (the successor to Brooklyn Union Elevated) than toward Interborough Rapid Transit Company, and he rarely missed an opportunity to slam it.[21]

Hylan found that his attacks on the transit companies improved his standing with the electorate. New Yorkers had not forgotten that transportion magnates such as Jay Gould and Russell Sage had bribed aldermen, countenanced poor train service, and sold worthless stock to investors.[22] In some important ways John F. Hylan expressed the needs of New Yorkers who had come to depend much more heavily on the subways in the 1920s. No census of subway riders was ever taken, and it is extremely difficult to gauge changes in passengers' social characteristics over time with any degree of precision; guesswork and the rule of thumb almost invariably take the place of reliable statistics. It nonetheless seems that the proportion of rapid transit riders who were poor and working class increased significantly in the 1920s as a result of the dual contract subways' extraordinary dispersal of the population to the outer boroughs. Many of the residents of working-class neighborhoods, including Little Italy and the Lower East Side, had walked to work; as they moved from these old Manhattan areas to outlying sections—Brooklyn's Bay Ridge and Canarsie and Queens' Astoria and Woodside—many relied on the subway to reach their jobs. For workers who earned only $4.00 to $8.00 dollars per day, a low-fare subway was important. With his spirited advocacy of the nickel fare, Mayor John F. Hylan aided poor New Yorkers. In voicing the needs of his constituents, in articulating and manipulating strong public resentments, Mayor Hylan made the political system work.[23]

Yet Mayor Hylan failed to fashion long-term solutions to transit's grave financial ills. He was always more interested in bashing the IRT and, particularly, the BRT on the front pages than seeking remedies for the subways' problems. His outlook on transit was negative and antagonistic, and he scorned the possibility of reaching a compromise with the private subway companies and public regulatory commissions.

For all of his personal disdain for the IRT and BRT, Mayor Hylan was acting within the framework of a distinctive political culture. Hylan was opportunistically filling a void caused by a more general political failure. The key elements of the subway's political culture were that the combination of public ownership and regulation and private operation of the subways encouraged competition rather than compromise, that strong popular pressures for a low fare and good passenger standards increased the friction between business and government, and that overriding resistance to public investment in rapid transit perpetuated the system's fiscal crisis. This fractious political culture led to paralysis; arriving at decisions was extremely difficult because many crosscurrents were at work, and the subway became imprisoned in a long, frustrating series of disputes that never produced lasting solutions.

Hylan received an opportunity to express his disdain early in his first term when the BRT experienced the worst disaster in New York rapid transit history. At 5:00 A.M. on Friday, November 1, 1918, the Brotherhood of Locomotive Engineers went on strike to protest the BRT's dismissal of twenty-nine motormen for unionizing. To break the union, the BRT kept the trains rolling and pressed supervisors, clerks, and other managers into service as motormen and conductors. One such scab was a twenty-three-year-old dispatcher named Edward Luciano who was not qualified to be a motorman. He had received only two and a half hours of training instead of the standard twenty days, and he had never operated a train alone before. But Luciano nonetheless put in a full shift as a motorman on the Culver line. Finishing his last Culver run late in the afternoon, Luciano was exhausted by the emotional strain of the strike as well as by the demands of this new and unfamiliar job. He was also reeling from the deadly influenza epidemic, which killed hundreds of thousands of people in 1918 and 1919. Luciano was not only recovering from a recent bout with the virus but was also mourning the death of one of his children, a recent victim of it. Tired and shaky, he wanted to go home when his shift ended. But BRT trainmaster Benjamin M. Brody, who was short of men, ordered him to take out a five-car Brighton elevated train. Luciano's first trip, inbound from Brighton Beach to the Park Row terminal across from city hall in Manhattan, was routine. His second trip was anything but routine. He started out shortly after 6:00 P.M., at the height of the evening rush hour. He crossed the Brooklyn Bridge and forgot to stop at the Sands Street station. Then he missed a switch at the Franklin Avenue junction, where the Fulton Street and Brighton els diverged, and mistakenly headed down the Fulton Street line. Realizing his error, Luciano reversed direction, returning to Franklin and switching onto the Brighton route. The stretch of track that led from the Franklin Avenue switch to the Malbone Street station was particularly treacherous. It consisted of a steep grade where the track descended below ground, a sharp S-curve, and a short tunnel that terminated in the station. Luciano's train roared down the hill shortly before 7:00 P.M., entering the curve going much faster than the posted speed limit of six miles per hour. Luciano, who later claimed that his brakes had failed, estimated his speed at about thirty miles per hour. Some eyewitnesses contended, however, that the train was going seventy miles per hour. The train derailed in the curve. The first car escaped relatively unscathed, but the next two cars hit the concrete tunnel so hard that the sound of the impact was heard a mile away. A newspaper reporter wrote that these "old wooden cars . . . crumbled like fruit cases when they struck the concrete wall of the Malbone Street tunnel."[24] The second car, number 80, and the third car, number 109, were both thirty-two-year-old wooden motorless trailers. The roof and the left-hand side of car number 80 was ripped open, and car number 109 was completely destroyed. Passengers were catapulted out of the two cars and onto the roadbed. Because the train kept moving after it left the rails, the wheels of the fourth and fifth cars ran over the riders who had landed on the tracks. Some were decapitated, some had their arms and legs sliced off. Other passengers who had been riding in the second and third cars were hurled against the tunnel wall. Still moving, the train ground them against the concrete in a horrifying "mill of death," leaving a shapeless mass of human flesh, wood splinters, and metal fragments.[25]

The train finally shuddered to a halt. Dazed, terrified survivors began picking their way out of the wreckage. One rider, eighteen-year-old Frieda Lee, said:

> I was knocked unconscious when the crash came. When I had regained my senses, I saw two men coming through the wreckage with lanterns. I could see many persons with their legs cut off. One woman had lost an arm and was begging for someone to kill her, saying the fire was coming. Women were hysterical. I was bruised. I heard one child calling for its mother.[26]

Sidney Weiss, an office boy for the *New York American* who was riding on the platform of the last car, told a reporter from that paper:

> Suddenly the lights went out and then came a crash. My nose was knocked against the rail, and I fell on the floor, which was wrecked. Some of those near me were thrown right out and underneath the train. I think one was killed. I got out very easily. With some of the men I helped women to get out of the car. It was all telescoped and broken up. All I could see of it was just a tangled mass of broken wood and twisted iron. I knew a lot of people were killed and badly hurt. Men and women were screaming, "Where's my sister?" "Where's my mother?" and so on.[27]

Policemen, firemen, and other people rushed to the station. They discovered that the tracks were strewn with jewelry, overcoats, fur pieces, handbags, mounds of wreckage, and dozens of dead bodies. The hunt for survivors was impeded by the darkness of the tunnel. Later, automobiles were driven to the edge of the open cut and their headlights trained on the tracks to provide light for the rescuers.

News of the disaster flashed through Brooklyn's neighborhoods. Hundreds of people went to Malbone Street that night, frantically seeking word about friends and relatives who might have been aboard the train. Crowds besieged the hospitals, too. Thousands milled around the station all the next day, either trying to find out about their loved ones or gawking at the scene of the disaster. At least ninety-three people died at Malbone Street, which remains the worst accident in the history of mass transit in the United States. Memories of the horror did not fade. The following year Malbone Street's name was changed to Empire Boulevard to create emotional distance from the tragedy. The crash pushed the Brooklyn Rapid Transit Company—already weakened by inflation—into bankruptcy, on December 31, 1918. When the railway emerged from receivership in March 1923, it was renamed the Brooklyn-Manhattan Transit Corporation (BMT) to erase the lingering stigma.[28]

The Brooklyn Rapid Transit Company was directly responsible for Malbone Street. Although Edward Luciano operated the death train negligently, the BRT was to blame for assigning this exhausted, untrained dispatcher to run the train in the first place. As Mayor Hylan charged, the BRT thus showed "reckless disregard for human life."[29]

Hylan reached Malbone Street late in the evening of November 1. The next day he invoked an obscure provision of the city charter that gave the mayor the power to sit as a judge in magistrate's court in criminal actions. Hylan then went to the Snyder Avenue courthouse in Flatbush and began hearing evidence about the accident. As subway historian Brian J. Cudahy noted, Hylan was mainly interested in embarrassing the BRT and acted like a "hanging judge."[30] Hylan's shortcoming, with Malbone Street and with transit affairs as a whole, was not that he voiced popular anger but rather that his rage was not accompanied by positive actions. He had no interest in constructive endeavors that might improve the lives of ordinary people, such as settling the BRT strike, trying to raise workers' wages, or upgrading railway safety. Instead, he was satisfied with venting his spleen and getting his name in the next day's headlines. There were no substantive results from his investigation.[31]

"Transit Manipulation and Transit Chicanery"

During his first term in office, John F. Hylan blocked regulatory efforts to restore the transit companies' ruined finances. After the war, the New York State Public Service Commission put together several plans to reorganize the railways' finances.[32] Hylan vigorously denounced these PSC proposals as a "cynical exercise" in "traction manipulation and traction chicanery" intended to bamboozle the people, save the bondholders, and raise the nickel fare.[33]

Hylan escalated his criticism of these bailout proposals in 1921 when a new agency, the New York State Transit Commission (TC), supplanted the Public Service Commission. Hylan objected to the Republican-spawned, upstate-influenced Transit Commission on party and home rule grounds. He also detested the TC's chairman, George McAneny, who of course was best known as the architect of the dual contracts of 1913. Hylan accused McAneny and the Transit Commission of conspiring with the Interborough and Brooklyn Rapid Transit companies and of plotting to open the city gates so that the subway executives could loot and sack New York "like the conquered cities of old."[34]

What gave some credence to Hylan's fantastic charges was the unanticipated results of the dual contracts. In these 1913 agreements George McAneny had tried to bind the two companies and the city together in a new era of cooperation that would put an end to the battles over monopoly and subway construction. The key to the dual contracts was a profit-sharing formula for disbursing each corporation's gross revenues: The company would first receive a preferential payment equal to the profits on its existing lines and the funds needed to retire its construction bonds, and then the municipality would recover its fixed charges, and the company and the city would divide the surplus.

The dual contracts, however, were drawn up in a different era. When the agreements were signed in 1913, the transit industry was healthy. The Interborough, for instance, was making profits of close to 20 percent. Consequently, the negotiators for the Board of Estimate and the Public Service Commission assumed that earnings would remain high. No one even imagined a situation in which subway revenues might prove inadequate to cover costs. In 1937 former Public Service Commissioner George V. S. Williams explained their thinking:

> The Interboro was earning and paying large dividends, had good credit and positively refused to enter into the contract except that an amount equal to their average annual earnings for two years was allowed that company out of annual operating earnings under Contract No. 3. The Brooklyn Company demanded the same terms. . . . So confident was everyone connected with the discussion that the question most considered was not "if" but when, a division of the profits would commence.[35]

What happened after World War I was that the IRT and BRT preferentials consumed nearly all of the subway receipts, not just a fraction. As late as 1922 the city had not collected any funds from the two companies. During the entire twenty-one years of its profit-sharing agreement with the Interborough Rapid Transit Company, from 1919 until 1940, the city took in only $2.1 million of the total IRT dual contract revenues of $306.9 million. Worse, the city did not obtain a single penny of the BRT's $207.3 million. Over the same period the city government accumulated a deficit of $278.7 million with the IRT and $182.7 million with the BRT, for a total loss of $461.4 million. Without these projected revenues, the municipality had to pay for its fixed subway costs—over $10 million annually—out of its general tax budget.

This financial reversal heightened press and popular resentment of the subway companies and gave Mayor Hylan ammunition for sniping at TC chairman George McAneny.[36] And although inflation was probably the factor most responsible for the altered circumstances of the city's rapid transit, Mayor Hylan did have a point. Much political power had been consolidated in so-called hidden or private governments. During the nineteenth century, private associations such as the Chamber of Commerce, City Club, and Citizens' Union had been active in public affairs, sometimes—as in the case of the Chamber of Commerce's domination of the Rapid Transit Commission—exercising governmental functions. The role of such hidden governments had changed during World War I when the federal government established the War Industries Board, the U.S. Shipping Board, the U.S. Railway Administration, and other agencies to supervise the mobilization of the private economy for the war. Dedicated to businesslike efficiency and determined to have decisions made outside electoral politics, these boards exercised the authority of the state and combined representatives of government, business, and labor unions. On the basis of the influential federal wartime model, a number of hidden governments were established in the 1920s on the federal, state, and local levels. In New York, for instance, semi-independent government corporations were created, such as the Port of New York Authority (formed in 1921 to reorder the region's ports and terminals), and private organizations were established, such as the Regional Plan Association (created in 1922 under the auspices of the Russell Sage Foundation to rationalize the physical metropolis). These specialized, professional organizations exercised great influence on New York's regional development.[37]

Hylan abhorred these hidden governments. He consistently opposed the consolidation of private decision-making. He fought the creation of the Port Authority and tried to abolish zoning codes. As a veteran of a local political club and as the self-styled champion of the common man, Hylan distrusted the anti-political thrust and professional orientation of these organizations. Having been elected mayor in a bitter contest centered on the loyalty of his German and Irish supporters to the federal government, and perhaps reacting to the surging nativism that ultimately culminated in the immigration restrictions of the National Origins Act of 1924, Hylan was sensitive to the centralization of power. As an urban populist who railed against the plutocracy, Hylan was suspicious of covert alliances between big business and government leaders.[38]

In his dispute with McAneny and the Transit Commission, Hylan put forth his own ideas for solving the traction crisis. He wanted the city to take over the subways, elevateds, and street railways and combine them into a single operating agency that would benefit from scale economies and more efficient public administration. Hylan depicted city operation as a panacea that would restore the railways to financial health, squeeze the water from the private securities, and save the nickel fare. Indeed, he expected the subways to earn a surplus that could be used for the construction of schools, hospitals, parks, and highways. Hylan never bothered to provide detailed financial estimates, but his program nonetheless seems completely unrealistic. Even if local government had been able to raise the hundreds of millions of dollars needed to buy the privately owned street and elevated railways, the publicly owned and privately operated subways could not have been acquired for any price for several years. Under the recapture conditions of the dual contracts, the city could not acquire any subway route until the new lines had been in operation for ten years; this date was 1925 for the IRT and 1926 for the BMT. More important, municipal operation probably could not resolve the underlying problem of declining transit income, especially since Hylan intended to keep the five-cent fare.[39] The famous American historian Charles A. Beard, who was an expert on municipal government, observed that public operation would not be able to make the railways solvent "by any scheme of reckoning known to man" unless the fundamental question of fare revenues was settled. "The city cannot coin money," Beard noted acidly. "The city cannot finance its municipal ownership program in the face of [banking and public] opinion that the financing is not on a firm basis."[40]

Nothing better symbolized John F. Hylan's politicization of rapid transit than his elevation of the nickel fare as perhaps the dominant electoral issue in New York City.[41] It was the centerpiece of his 1921 reelection campaign. "This five-cent fare," he declared, "is the cornerstone of the edifice which we call New York City." Lauding the nickel fare as "the basis on which New York City spread out and on which the suburbs have developed," Hylan argued that the low ticket cost was a "property right" for the middle-class and working-class homeowners who had moved to the new neighborhoods on the outskirts. Because of its significance to the masses, the mayor considered the preservation of the nickel fare for the duration of the dual contracts "as sacred and binding as any contracts ever drawn in the history of financial transactions the world over."[42] "My policy," Hylan later affirmed, drawing an explicit link between the welfare of ordinary citizens and the low price of a subway ride, "has been the preservation of democracy and the retention of the five-cent fare."[43]

John F. Hylan won a landslide victory in November 1921, capturing 65 percent of the ballots. Most observers credited his triumph to his championing of the nickel fare,[44] which was transformed into an article of faith so sacred that no mayor risked raising it until 1948, when it was increased to a dime.[45]

The five-cent fare was not John F. Hylan's only contribution to the subway. After his 1921 reelection he focused on another important issue: the construction of new subways. During his second term, the mayor brought the same strident rhetoric and anti-corporate crusading to rapid transit planning that he had already brought to the nickel fare and fiscal reorganization. The result was that subway expansion became more competitive and conflict-ridden. This politicization of rapid transit construction ironically contributed to the shift from rapid transit to the automobile.

The Road Not Taken: Rapid Transit and Visions of the Future City

In the early 1920s, John F. Hylan had no legal authority to build subways. Rapid transit planning was the province of state regulatory boards: the Office of Transit Construction Commissioner (1919–21) and then the Transit Commission (1921–24).

A key figure with the Transit Construction Commissioner's office was Daniel L. Turner, an engineer who was born in Portsmouth, Virginia, in 1869 and educated at Rensselaer Polytechnic Institute in Troy, New York. In 1900, Turner landed a junior position on William Barclay Parsons' Rapid Transit Commission engineering staff. After the Contract No. 1 subway was finished, Turner decided to make a career of subway engineering. He became chief engineer of the Public Service Commission in 1917; two years later the Office of Transit Construction Commissioner acquired the PSC's power to build subways, and Turner became its chief engineer. A dynamic, gregarious man who belonged to numerous professional and social organizations and a skillful writer who published pieces in engineering periodicals as well as in the popular press, Turner was effective at promoting rapid transit's potential to reorder the physical city.[46]

Like many other engineers and planners, Turner accepted the main tenets of the dual contracts of 1913.[47] He perceived of rapid transit as a planning tool for the rationalization of urban development and the dispersal of the population to the outskirts, and he thought of the new routes as extensions of the IRT and BRT networks. The continuation of these ideas constituted a dual contract tradition in subway planning that provided an important alternative vision of the future of New York City.

Although Turner approved of the dual contract system's goals, he complained that its decisions were not sufficiently rationalized. He thought that George McAneny had been so preoccupied with the immediate squabbles between the government and the subway companies that the dual system neglected city planning's higher aims of systemization and integration. For instance, Turner felt that McAneny missed an opportunity to create a unified subway. Turner complained that McAneny had been so eager to compromise with the Interborough and Brooklyn companies that he stuck New York City with two separate subway networks. Turner also thought that McAneny should have prepared a long-range construction program to govern rapid transit planning for the next thirty or forty years. In fact, Turner's perfectionism was a serious weakness; he lacked the knowledge of business and the readiness to compromise that had characterized George McAneny's negotiation of the dual contracts.[48]

In July 1920, Turner issued a plan for the TCC that embodied his ideas about subway design. His proposal called for the construction of 830 track miles in three main parts. First, Turner would construct two new trunk lines in Manhattan and the Bronx: an eight-track Amsterdam Avenue–Eighth Avenue route on the west side and a six-track Madison Avenue subway on the east side. Second, he would lengthen some existing BRT and IRT subways to better integrate the rapid transit network and provide access to the outlying districts. For instance, he wanted to inaugurate crosstown service between Queens and Brooklyn by sending the BRT's Astoria line south through Long Island City, Greenpoint, and Williamsburg to a connection with its Brighton elevated in downtown Brooklyn. He also sought to extend the BRT's Fourth Avenue subway by digging a two-track tunnel from Bay Ridge, Brooklyn, under the Narrows to Staten Island; this tunnel would open the city's most thinly settled borough for residential development. Third, Turner would relieve the severe traffic that afflicted midtown Manhattan by building transit lines from river to river, on Fourteenth, Forty-second, and Fifty-seventh streets.[49]

In formulating this proposal, Daniel L. Turner made critical assumptions about New York City's future. He thought that rapid transit would continue to be the chief mode of transportation and that modern subways offering fast service would crisscross the city. Because of rapid transit's significance, New York would remain centralized and densely populated. The built-up area would still be confined to the city limits, and the five boroughs would hold nearly all of the region's inhabitants, commercial businesses, and factories.[50]

As a result Turner was convinced that New York City's population and mass transit patronage would continue to increase as explosively as in the past. He expected that by 1945 the city would contain 9.5 million people and its subways and els would carry 5 billion passengers. For this reason Turner thought his TCC plan would soon be inadequate. Even though it would double New York's rapid transit mileage, he was afraid his new lines would be overwhelmed by the mounting traffic. He contended that the only way to avoid this peril was by opening more and more subways. "In order to keep pace with the enormous traffic growth," he declared, "the city must build more transit facilities—then more again—and still more again—and must keep on doing this continually."[51]

To do so, Turner urged New York City to adopt sound planning principles and build rapid transit continually.[52] What Turner failed to see, of course, was the growing importance of the automobile, which became the progenitor of the future city. Many Americans came to the conclusion that the automobile represented the latest advance in the march of technological progress and that railways should be condemned to the scrap heap.[53] By the 1920s even many urban planners favored the automobile over mass transit.[54]

Perhaps the most significant planning document in the city's history emerged in the late 1920s. The *Regional Plan of New York and Its Environs* (1929–1931) recommended a balance of vehicular and subway improvements. Probably its best-known contribution was its proposal for a vast arterial and circumferential highway system, with bridges and tunnels running across the Hudson River and the East River, and parkways going through the city and suburbs.[55]

But *Regional Plan* did not neglect rapid transit. The chief author of *Regional Plan*'s transit program was Daniel L. Turner, who had entered private practice after retiring from the Office of Transit Construction Commissioner in 1921. For his program Turner drew on ideas he had developed in his 1920 TCC plan; indeed, the later design may be considered the ultimate expression of the dual contract tradition of subway planning. Turner called for the construction of a gigantic system of 297 miles of track at a cost of over $1 billion. He wanted to upgrade New York City's rapid transit lines by replacing old elevateds with new subways, building a new subway on the Upper East Side, and digging tunnels below the East River and the Narrows. More important, Turner sought to send the city's railways beyond its boundaries for the first time, forming a system that was genuinely regional in scope and would eliminate political boundaries as transportation barriers. He envisioned a splendid network of "suburban rapid transit" lines going to the New York City–Nassau County border and penetrating New Jersey's Passaic, Essex, Bergen, and Union counties.[56]

These recommendations probably could have been implemented. Turner's plan did not entail significant design or mechanical advances and probably could have been built with existing technology. This subway proposal would have been expensive, but it was no more prohibitive than the remarkable chain of vehicular improvements undertaken during the 1920s and 1930s. Hundreds of millions of dollars were poured into the George Washington Bridge, Lincoln Tunnel, Holland Tunnel, Bronx-Whitestone Bridge, Triborough Bridge, Queens Midtown Tunnel, Interborough Parkway, Grand Central Parkway, and other projects. In view of the enormous sums allocated to highways during this period, it seems likely that at least part of Turner's proposal could have been carried out. The rapid transit program of the *Regional Plan*s suggested an alternative to the one-sided orientation toward highways that evolved in the 1930s and afterward. There was a way for subway construction to continue on a large scale after 1940 instead of coming to a virtual halt as it did.[57]

The failure to build these subways was primarily due to mass transit's flawed political culture. Daniel L. Turner himself grasped the importance of this issue in his *Regional Plan* proposal that the rapid transit lines be controlled by a type of government called a public authority or a special district. As a semi-independent corporation staffed by professional experts and dedicated to efficiency, this public authority would be free from the electoral pressures and democratic controls that stopped the executive and legislative branches from taking decisive action. Unhampered by the narrow geographical jurisdictions and parochial outlooks of local municipalities, the public authority could function on the broad regional scale needed to realize Turner's vision.[58]

Turner's emphasis on this modern public authority was significant precisely because New York's subways had a very different political framework. Its subways and els were managed by private corporations and regulated by public commissions in a nineteenth-century pattern of industrial organization that was conducive to competition rather than coordination. The fundamental problem stemmed not from the type of government structure that controlled the subways—regulatory commission, municipal agency, or public authority—but rather from the general political values, practices, and institutional relationships that prevailed. With his emphasis on the flexibility and independence of the public authority, Turner exposed the faultlines in the subways' political culture: the conflicts between the subway companies and government leaders, the stress on business profits, the unwillingness to invest public money in rapid transit, and the persistence of a New York City orientation instead of the adoption of a wider regional focus. As the confrontation over the nickel fare demonstrated, the Hylan administration, the Transit Commission, the Interborough Transit Company, and the Brooklyn-Manhattan Transit Corporation were consumed by intractable disputes that constricted their vision and made cooperation impossible. Transit leaders were incapable of the entrepreneurial energy and metropolitan outlook demanded by the *Regional Plan*.[59]

Public authorities such as the Port Authority and Robert Moses' Triborough Bridge and Tunnel Authority did have a great impact on New York's regional development in the 1920s and 1930s, but they built highways, not subways. The automobiles were not impeded by transit's sclerotic political structure, and its reputation for being nonpolitical and technologically modern facilitated the approval of capital projects. In the 1930s and afterward, these authorities opened magnificent bridges, tunnels, and expressways that radically changed metropolitan New York.[60]

As Kenneth T. Jackson observed in *The Crabgrass Frontier* (1985), the adoption of the private automobile and the construction of highways set the stage for suburbanization. Together with such factors as the large supply of cheap land, personal affluence, rapid population growth, pervasive racial fears, and government subsidies to homeowners through low-interest federal loans and the tax system, Jackson credited motor vehicles for bringing about the radical decentralization of New York. The region became a decentralized, low-density "spread city" that sprawled across thousands of acres as jobs and housing followed the superhighways out of New York City.[61]

Rapid transit became a foreign presence in most parts of this metropolis. Subways remained important in New York City, however. The municipal government did embark on a significant subway construction program in the 1920s, but Mayor Hylan made sure that the new network bore his imprint.

The Independent Subway System

Mayor John F. Hylan had his own vision of the subways. In September 1922 he published a thirty-page pamphlet entitled *Mayor Hylan's Plan for Real Rapid Transit,* which argued that subways should be "planned, built, and operated to accommodate the transportation needs of the people . . . and not solely for the financial advantage of the operating companies or their officials."[62] Hylan envisioned an independent system owned and operated by the municipality that would compete directly with the IRT and BMT for riders. By giving passengers an alternative to the "grasping transportation monopolies," moreover, a municipal subway would put pressure on the IRT and BMT to sell their railways to the city at a good price.[63] When the acquired IRT and BMT subways were combined with its new municipal subway, New York could boast a citywide, city-operated network that would take it "to the pinnacle of social, civic and commercial success."[64]

The greatest obstacle to Hylan's vision of an independent subway was the fact that the Transit Commission possessed legal authority over rapid transit development. Only four months earlier, in May 1922, TC chairman George McAneny had issued a subway proposal of his own. Like Daniel L. Turner's blueprints for the TCC and the *Regional Plan,* this proposal belonged to the dual contract school of using rapid transit to reorder the urban environment and designing the new lines as extensions of the IRT and BMT subways. McAneny incorporated many of Turner's TCC routes, including the Staten Island tunnel, the Amsterdam–Eighth Avenue line, and the Forty-second Street crosstown route.[65]

Hylan protested that the Transit Commission laid out its routes so only the Interborough and Brooklyn companies could operate them. To George McAneny this integrated design made sense as a means of extending transit to the outskirts and relocating people from the center. But Hylan did not share McAneny's concern for the urban landscape; to the contrary, the mayor was primarily worried about the balance of power between the public and private sectors. Hylan regarded the Transit Commission's goal of coordinating the IRT and BMT railways as evidence of McAneny's "subtle and sinister" conspiracy to enrich the companies. Just as the "artful and cunning" McAneny had "served the traction ring faithfully and well when he put over the dual contracts," so McAneny's "traction-created and traction-controlled state agency" was now hatching "the biggest traction grab ever attempted."[66]

Hylan wanted to wrest control of subway planning from the Transit Commission. He received valuable support from New York State's leading Democrat, Al Smith. First elected governor in November 1918 and then defeated in his bid for a second two-year term, Smith was trying to regain his seat in 1922. Smith and Hylan were unlikely allies. They not only disliked each other personally, but they constantly jockeyed for power within the Democratic party.[67]

But Tammany boss Charles F. Murphy usually managed to keep Smith's and Hylan's antagonism in check, however, and the two men were sometimes able to cooperate. Even though Smith did not share Hylan's lust for municipal operation or his hatred of the IRT and BMT, they both opposed state transit regulation of the subways as a violation of municipal home rule and as a GOP encroachment on Democratic prerogatives. Moreover, Al Smith sensed that the Transit Commission would make a good campaign issue. During the 1922 gubernatorial campaign, Smith blamed the Transit Commission for delaying subway construction and for plotting a fare hike. He promised to abolish the TC and replace it with a municipal board.[68]

Smith won the 1922 election, but he had trouble redeeming his campaign pledge. For over a year the Republican majority in the Assembly blocked the governor's efforts to kill the TC. Then, in February 1924, Republican Assembly Speaker H. Edmund Machold and Majority Leader Simon L. Adler fashioned a compromise measure, called the Adler bill, designed to save the TC by dividing the jurisdiction over city transit between the state and municipal governments. Machold and Adler wanted to protect New York State Transit Commission's regulation of existing lines in order to safeguard GOP patronage jobs and shield the surface railways from potentially hostile municipal regulation. In return for sparing the TC, Machold and Adler offered to transfer the power to develop rapid transit to a new bureau, the New York City Board of Transportation, which would report directly to Mayor Hylan and be able to build and operate its own subways.[69]

In addition, the bill contained a provision, called the self-sustaining fare, intended to guarantee the subway's fiscal soundness; this was added in response to criticism of the original bill by business leaders. This provision set the fare at a nickel for the first three years of municipal operation; afterward the fare would have to cover the subway's operating costs and capital debt. The purpose of the self-sustaining fare was to ensure that operating losses were paid by the riders through higher prices at the turnstiles rather than by realtors and other businessmen through higher property taxes.

In April 1924, Governor Smith announced his support of the compromise bill. Hylan reluctantly decided to endorse the measure, too, for the sake of obtaining municipal operation of rapid transit. Smith signed the law on May 2, 1924.[70]

The new Board of Transportation took office on July 1, 1924, with a hard-nosed Tammany veteran, John H. Delaney, as its chairman. Delaney began releasing his proposal for city-run routes in December 1924. The IND had two main purposes: to provide faster service in established business districts and older residential neighborhoods, and to replace the elevated railways with newer underground railways. In its final form the new underground railway, the Independent Subway System (IND), would have seven major routes and some 190 miles of track. (See Map 4, the Independent Subway System, 1940.) The first route, the Eighth Avenue, would run across the East River from Brooklyn Heights to lower Manhattan and then up Church Street and Eighth Avenue to 145th Street in Harlem, where it would split into two branches. One branch would follow St. Nicholas Avenue and Broadway to 207th Street in northern Manhat-

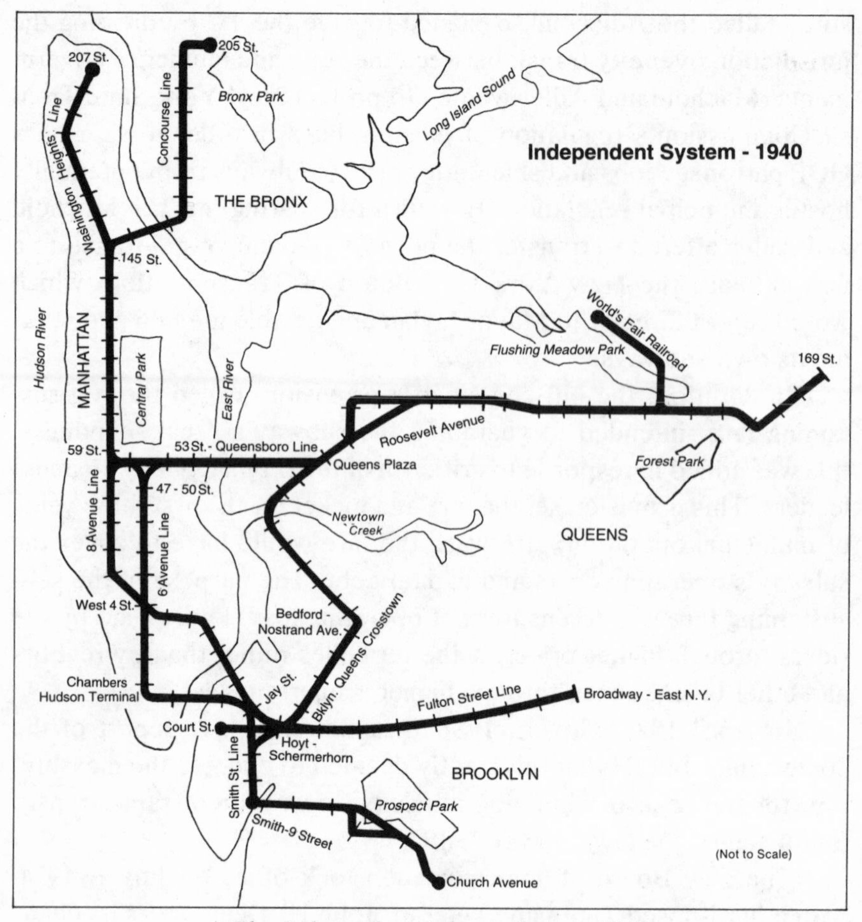

4. The Independent Subway System, 1940.

tan, and the other would head up the Grand Concourse to 205th Street in the Bronx. The second route, the Sixth Avenue, would run under Sixth Avenue from Fifty-third Street to Greenwich Village. There it would connect with a third IND line, the Houston Street, which would cut through the Lower East Side, pass under the East River, and end at Brooklyn's Jay Street. The fourth IND subway would follow Fulton Street from downtown Brooklyn to East New York. The fifth route, the Smith Street line, would go south from downtown Brooklyn to Red Hook, skirt the edge of Prospect Park, and terminate in the Kensington neighborhood. A sixth route, the

Queensboro, would provide more service for fast-growing Queens, going from midtown Manhattan to Queens Boulevard and 169th Street in Jamaica. The seventh route would let subway riders travel between Brooklyn and Queens without having to go through Manhattan first; this crosstown line would wind from Bridge Plaza in Long Island City, Queens, to Fulton Street in the Brooklyn business district.

The locational pattern of the Independent Subway System routes was substantially different from that of earlier rapid transit lines. (See Map 5, Development of New York City Rapid Transit.) Unlike the elevated railways, the Contract No. 1 IRT, and the dual contract subways of 1913, the IND lines were primarily confined to the built-up districts. Of the seven IND routes, only the Queensboro stretched beyond the urbanized zone to sparsely settled territory on the outskirts. The reason for this pattern was that the Independent was a separate system with no existing source of revenue; like the ill-fated Triborough plan of 1908, the IND routes had to be concentrated in business districts or heavily populated neighborhoods that already had substantial passenger traffic in order to avoid huge operating losses. To be sure, the IND would accomplish the important goal of furnishing the older areas with faster service. But Hylan's goal of creating an independent subway that could compete with the Interborough and the Brooklyn-Manhattan would thus severely limit the spatial impact of the new municipal system; except for the Queensboro line and the Washington Heights and Grand Concourse divisions of the Eighth Avenue routes, the IND subways would not be able to spawn new neighborhoods or redistribute the population to the outlying sections.[71]

The IND constituted an important break in city transit history. For the past ninety years, ever since the first horse car had trundled up the Bowery on November 26, 1832, mass transit had been essential to the city's outward expansion. During the nineteenth century, first the horse railways and then the elevated railways had stimulated Manhattan's march uptown. The 1904 subway had sparked the residential settlement of northern Manhattan and the Bronx, while the dual contract subways had dispersed the population to the outer boroughs. But Hylan's Independent System brought an end to rapid

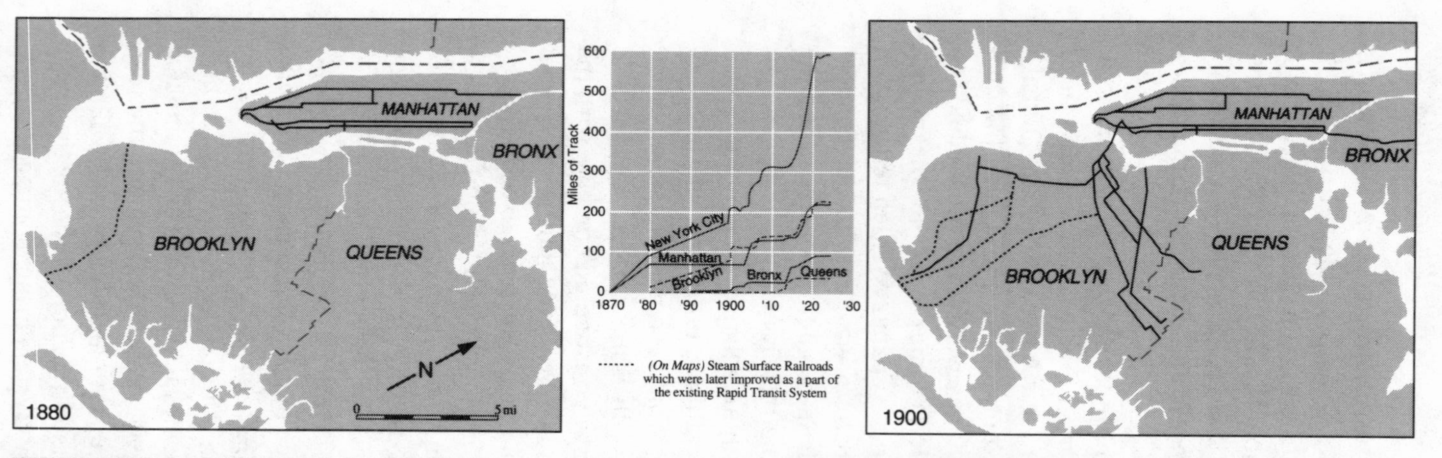

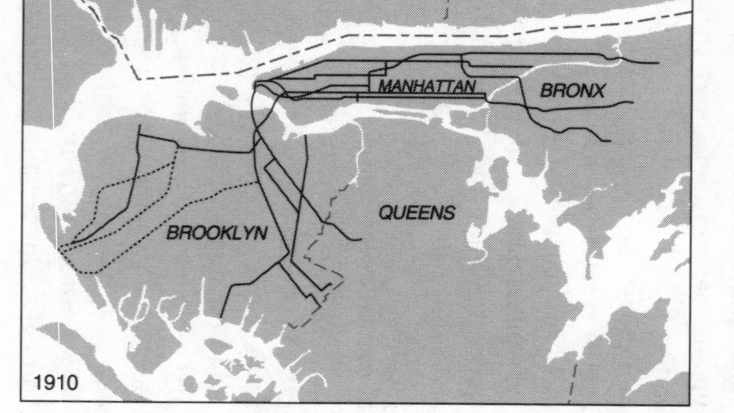

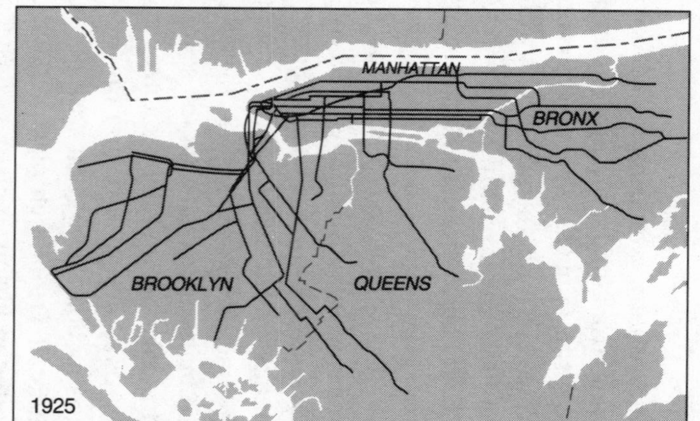

5. *Development of New York City Rapid Transit.*

transit's role as a spearhead of expansion. This critical change stemmed from Hylan's politicization of the conflict between government and business. Hylan put so much stress on competition with the IRT and the BMT that his new subway simply could not be extended to the new territory on the outskirts. In fostering direct municipal competition with the private subway companies, the mayor reinforced the shift from rapid transit to the automobile as the catalyst of residential development. At the same time that modern political structures such as the Port Authority were beginning to build highways across the metropolis, Mayor Hylan ensnarled rapid transit more tightly in a web of public controversy, intergovernmental conflict, and public-private disputes.[72]

But the IND would have an important effect on the city's landscape in another way: the replacement of elevated railways. Several IND subways would run along elevated routes: the Sixth Avenue IND would go below the Interborough's Sixth Avenue elevated, the Fulton Street IND would go under the BMT's Fulton Street el, and the Eighth Avenue IND would go within a block of the Interborough's Ninth Avenue el. This was deliberate. By the 1920s many New Yorkers spurned the els as fossils from the grimy industrial city of the nineteenth-century. Made obsolete by the introduction of the high-speed, high-capacity subways, the elevateds were considered old and slow and too unsightly to merit a place in the modern metropolis. The els were now perceived as traffic obstacles because their bulky iron pillars impeded cars, trucks, and buses. Transportation officials wanted to accommodate motor vehicles by segregating the competing transport modes on different levels. They wanted to reserve the surface for motor vehicles, replacing the els with subways and moving rapid transit passengers below ground. Mayor Hylan also saw the IND as a weapon in his war on the private subway companies. By diverting passengers from the IRT and BMT els to the Independent, he hoped to compel the corporations to agree to reorganization on his terms.[73]

Many businessmen who owned property along subway routes endorsed the IND because it would raise their land values. Boosters of Sixth Avenue, for example, believed the Independent would transform their thoroughfare into a prosperous office center. In the 1920s,

Sixth Avenue was a grubby collection of four- and-five-story brick tenements whose storefronts were occupied by cheap restaurants, furriers, barbershops, and drug companies.[74] "I think the plan is all right," commented Charles M. Dutcher, a bank president and the head of the Sixth Avenue Property Owners Association. "It gives Sixth Avenue the transit facilities it needs." Representing a northern Manhattan taxpayers' group, Meyer C. Goldman praised the Eighth Avenue subway for ensuring that "the people of Washington Heights will get shortly what they have been fighting for in the way of subway relief."[75] But the Independent received relatively little support from local businessmen in outlying areas because its routes were concentrated in built-up districts. Altogether, realtors were not as enthusiastic about the IND as they had been about the original IRT and the dual system.

The IND helped estrange realtors from the subway. Developers had boosted previous rapid transit lines in order to raise property values and settle outlying areas. Citywide promoters such as William Steinway, Charles T. Barney, and IRT President August Belmont, as well as local groups such as the Washington Heights Taxpayers Association (in Manhattan), Bensonhurst and Bath Beach Subway Association (in Brooklyn), and Edward A. McDougall's Queensboro Corporation (in Jackson Heights) had thus been powerful advocates for transit construction. But rapid transit became less attractive to realtors in the 1920s, in part because of Mayor Hylan's decision to concentrate the IND in the built-up area, and in part because of the automobile's ability to open much wider areas for settlement. Because the real estate industry had been perhaps the subway's most important constituency, its alienation had important political consequences. Beginning in the 1920s developers no longer lobbied as forcefully as they had earlier for new subway routes. And beginning in the 1940s realtors sought to cut municipal funding of subways in order to lower property taxes.

The IND's staunchest opponents were the IRT and BMT executives, who feared potentially ruinous competition with the municipal subway. Addressing the Men's Club of Congregation Beth Elohin in January 1925, BMT vice-president Travis H. Whitney assailed the IND as a boondoggle that confirmed Hylan's shortsightedness. According to Whitney, "the city administration has done nothing constructive toward providing comprehensive expansion of transit facilities" and prevented the BMT from "helping the city to secure more transit." If Mayor Hylan simply agreed to raise the confiscatory nickel fare, Whitney maintained, the reinvigorated BMT would be strong enough to build the Brooklyn-Queens crosstown, the Central Park West extension, and other new lines. Going ahead with the IND would benefit nobody, Whitney argued, since it would leave the city with three weak subways rather than two strong ones.[76] Whitney's boss, BMT chairman Gerhard M. Dahl, charged that the IND was so flawed, the Board of Transportation would either have to levy an eight- or ten-cent fare to break even or run annual operating losses as high as $172 million. Dahl predicted that the IND would bring the City of New York "to financial grief upon the rocks of unsound and impractical economics."[77]

Big business organizations such as the Chamber of Commerce and the Merchants' Association condemned the IND plan, too. They claimed that subways could be built more cheaply as extensions of the BMT and IRT networks than as a separate system. According to the Merchants' Association, the extension of the IRT and BMT would cost only $300 million, compared to $500 million for the Independent.[78]

To the Merchants' Association and the Chamber of Commerce, municipal subway operation was an invitation for elected officials to increase taxes and destroy the city credit. Politicians would do anything—award sweetheart deals to friendly vendors, pad the payroll with unnecessary employees—to curry favor and get reelected. In a speech to the United Real Estate Owners' Association in March 1925, State Tax Commissioner Martin Saxe declared that city officeholders coveted the IND "because they know that they can make themselves popular by giving away your money."[79]

This barrage of criticism did not faze Mayor Hylan, who asserted that the IND would mean "the end of traction rascality and threats of a 9 or 10¢ fare."[80] Hylan was certain that the IND would be approved. Because the Independent Subway entailed government ownership and operation rather than the more intricate partnership of public ownership and private operation that characterized earlier subways, the procedures for ratifying the IND were straightforward. The Adler law had established a two-step process for certifying a municipal operation: first, the Board of Transportation would adopt the routes, together with a statement of anticipated equipment, construction, and operating costs; second, the Board of Estimate and Apportionment would assent to the Transportation Board's routes and financial projections. The Board of Transportation submitted a comprehensive plan to the Board of Estimate on May 26, 1925. All of the routes in Manhattan, the Bronx, and Queens were accepted, but the Board of Estimate rejected some lines in Brooklyn. These routes were subsequently revised, and the Board of Estimate approved them in July 1927. The city stated that the IND would cost $674 million to build and equip and that it planned to raise that sum by issuing four- and fifty-year bonds.[81]

Construction had already begun. On the afternoon of March 14, 1925, Mayor Hylan broke ground for the first IND subway, the Washington Heights branch of the Eighth Avenue subway. To underscore the mayor's point that the IND was the "people's subway," the groundbreaking ceremony was held in a public park—John Hancock Park at the corner of St. Nicholas Avenue and 123rd Street in Harlem—and it was broadcast over radio station WNYC and filmed by newsreel cameras. Before lifting a silver pickaxe to break the sod, Hylan hailed the Independent for delivering New York from the grip of the traction ring:

> It means the beginning of the emancipation of the people of the City of New York from the serfdom inflicted upon them by the most powerful financial and traction dictatorship ever encountered. It means the first effective blow to the million-dollar traction conspiracy which shows its ugly head now and then in individual or concerted public statements by traction officials, members of the State Transit Commission, and spokesmen of so-called civic groups.[82]

The first IND subway, a section of the Eighth Avenue route, entered service on September 10, 1932. The last IND subway, the much-delayed Sixth Avenue route, opened on December 14, 1940.[83]

The decision to build the IND was a mixed blessing. Its construction was particularly noteworthy for the failure to implement

Daniel L. Turner's compelling vision of regional rapid transit and the critical concession to the automobile of mass transit's historic role as a catalyst of metropolitan expansion. Still, the IND did constitute an important addition to New York's subways. The IND routes tied New York's existing neighborhoods and business centers more closely together, providing better transportation for growing areas such as mid-Manhattan.

9

THE PEOPLE'S SUBWAY, THE NICKEL FARE, AND UNIFICATION

A Golden Age

For subway passengers the 1920s and 1930s were a golden age. The low nickel fare made rapid transit a bargain and enabled poor New Yorkers to take full advantage of the city. Thickly crowded with people of different classes and ethnic groups, the subways became known as a place of excitement and rich urban color that embodied both New York's grit and its endless possibilities.

Although the Contract No. 1 IRT subway had been perceived as an emblem of technological modernity in the beginning, underground rapid transit eventually grew older and became so commonplace that most New Yorkers were no longer excited by it. By the end of World War I, newer and faster transport such as the automobile and the airplane had replaced the subway as a symbol of speed, power, and freedom. But the subway nonetheless continued to have a hold on the popular imagination; now, however, its appeal stemmed from the human character of its crowds, not from its complicated, impersonal machinery.[1]

Artists helped formulate the new image of the "people's subway." Painters and printmakers such as John Sloan, Reginald Marsh, Isac Friedlander, and Isabel Bishop, who belonged to the so-called Ashcan School, made rapid transit one of their favorite subjects. Rejecting the more formal, restrained artistic traditions of the nineteenth century, they thought the subways and els—along with amusement parks, dance halls, bars, nightclubs, and similar attractions—revealed the emotional exuberance and social diversity of modern life. With their bustling crowds and grimy cars and stations, the underground and elevated railways communicated authenticity. As essayist Christopher Morley put it, the harassed riders who were packed into the subways somehow represented the "Truth" about humanity.[2]

For these artists the subway was a theater, a spectacle that presented a dizzying array of emotional encounters among male and female strangers of different ages, classes, and races. Marsh, Sloan, Friedlander, and the other artists knew that the subway was not an ideal world and that close physical contact among so many people from such diverse backgrounds often produced discomfort, alienation, and social tensions. Yet one emotion that was missing from their images of the subways was fear. In his 1939 etching *3 A.M.*, Isac Friedlander depicted ten riders: an amorous couple locked in a passionate embrace, seven passengers asleep or nodding off, and a rider engrossed in his newspaper. Even though it was the middle of the night, nobody on this IRT Lexington Avenue local seemed to be worried about his safety. In *Subway,* Harlem Renaissance painter Palmer Hayden portrayed a group of black and white passengers crammed together during rush hour who felt no sense of apprehension. In his 1934 wood engraving *Subway,* Fritz Eichenberg conveyed a similar impression of riders traveling on a Broadway–Seventh Avenue train late at night. Despite the fact that the passengers were divided fairly evenly between whites and blacks and between men and women, nearly everyone was fast asleep. This absence of fear was an important ingredient in the idea of the people's subway; in order to coexist with their fellow passengers and see subway riding as an adventure, New Yorkers had to feel safe underground.[3]

The subway also became a popular setting for movies and plays in the 1920s and 1930s. From a now-forgotten 1926 romantic comedy, *Subway Sadie,* to Warner Brothers' 1934 musical comedy *Dames,* which starred Dick Powell and Ruby Keeler, to William Saroyan's 1940 play *Subway Circus,* most plays and films used the subway setting to throw into greater relief a story of people falling in love or pursuing some personal dream. In *Subway Circus* and *Dames,* a subway car became a fantastic dreamworld that freed riders from the daily grind and relieved them of their anxieties; the train's noise, lights, and motion signaled that this fantasy was more exciting and fulfilling than reality itself. The subway lent dignity and meaning to the lives of ordinary people, especially during the great depression.[4]

This popular culture image of a people's subway corresponded fairly closely with reality. In the 1920s and 1930s, New York's subway was generally cheap, safe, reliable, and accessible. Providing speedy transportation to most parts of New York City, the subway was used by rich and poor, natives and immigrants. Indeed, due to the dual contracts' marked success in moving working-class residents from Manhattan to the outer boroughs, poorer New Yorkers appear to have constituted a larger share of total subway passengers in the 1920s and 1930s than before.

The key to the people's subway was the nickel fare, which allowed most people to use the system; for some, however, the five-cent rate still constituted a barrier. The low cost of a subway ride was a bargain, but a nickel was still a lot of money, particularly during the depression. For instance, Edward and Rosa Kaufman were German Jews who fled from Nazi persecution in their native Hamburg to New York City in 1939. Although Edward landed a job selling cheese at the Dutch pavilion of the 1939–40 World's Fair and Rosa found work in a garment factory, these refugees had left most of their belongings in Germany and were almost penniless. Too proud to accept charity, the Kaufmans survived by economizing. They regarded the five-cent subway fare an unaffordable luxury and often walked forty or fifty blocks from their apartment on West 107th Street in Manhattan to avoid paying for the IRT.[5]

A fare increase would have taken money from workers' pockets. Although no surveys of the social impact of a higher fare appear to have been made in the 1930s, a report that assessed the impact of a 7½-cent fare on a typical working-class family was prepared in 1943. According to this survey, if one member of a family that earned

$1,500 commuted to work six days a week, the cost would amount to $15.60, or 1 percent of its gross income. If two members of the family worked outside the home, the cost would have been $31.20, or 2 percent of its income. This was a lot of money for a poor family.[6]

But the nickel fare was low enough for millions to afford. It permitted poor New Yorkers to exploit the city's opportunities, despite their poverty. In his novel, *City Boy* (1948), a semi-autobiographical account that echoed his own memories of growing up in a Jewish neighborhood in the Bronx, Herman Wouk treated the subway almost lovingly as a source of adventure and knowledge. To Wouk the subway was an extension of the neighborhood and a mediator between the private life of the family and the block and the public sphere of the larger city.[7]

Many other New Yorkers looked upon the subway as a door to the larger city: classes at City College, musical concerts at Town Hall or Lewisohn Stadium, trips to the 1939–40 World's Fair. The subway was especially important to people who grew up in the remote, more parochial neighborhoods of the outer boroughs. Alfred Kazin, who later described himself as a "Brooklyn hick" from Brownsville, took the IRT and BMT to explore the city for himself. "There was a sense of discovery" about riding the subway, Kazin recalled, the possibility of "getting away to anything you pleased for a nickel."[8] Kazin commuted to City College at 138th Street and Convent Avenue in upper Manhattan and sometimes read Edward Gibbon on his long trips. Playwright Arthur Miller reportedly read *War and Peace* in the subway. Another City College student, Jerome Karle, spent three hours a day in the IRT and BMT, journeying from his family's apartment in Coney Island to the campus. Years later, after winning the Nobel Prize in chemistry in 1985, Karle remembered doing his homework in the subway and commented wryly, "I learned how to study under very odd circumstances."[9] In 1986, William Schuman, a composer and a former president of the Julliard School and of Lincoln Center, credited the subway with educating him about New York's lively arts culture. Schuman had very little money as a student in the 1930s, but he could still afford to attend plays, band concerts, the New York Philharmonic, and modern dance performances. He explained it this way:

> I saw many, many plays. The theater scene in New York was incredibly rich in those days. . . . What's more—and especially important for a young person of modest means—it was all affordable. My friends and I would take a five-cent subway ride to Times Square, have a marvelous lunch at Schrafft's for thirty-five cents, and buy a ticket for fifty cents—choosing from among fifty or sixty plays! With the whole day costing less than a dollar, we could do it every week—and we did.[10]

Native-born New Yorkers were not alone in their fondness for the subway. After graduating from the University of North Carolina in 1935 with a journalism degree, Vermont Royster packed up and moved to Manhattan, determined to try his luck despite the depression. Arriving full of dreams and eager to advance as a reporter, Royster was thrilled simply to be living in New York. Night after night he left his job, walked to the closest subway station, and went exploring—wandering around Greenwich Village or seeing the Algonquin's famous roundtable or going nightclubbing in Harlem. In his 1983 memoir, Royster, who by then had retired as editor in chief of the *Wall Street Journal,* wrote as follows:

> At twenty-one you're not inclined to get off work, eat dinner, and go to bed. Some nights I would go to a local movie house, most of which ran double features. Other nights to a neighborhood bar where you could nurse a ten-cent beer and watch the comings and goings. Many times, though, I would take the subway, that marvelous bargain for a nickel, and get off at some random stop to walk the Village or the Bowery. Sometimes I would spend an evening just walking around Times Square.[11]

A few observers loathed the subways, however, and mistakenly associated them with the poor. According to one account, Edna Woolman Chase, the longtime chief editor of *Vogue,* was shocked to learn that one of her editors had tried to commit suicide by jumping in front of a subway train. Chase reportedly summoned the other editors to her office and told them that such declassé behavior would not do. "If we must [kill ourselves]," Chase admonished, "we take sleeping pills."[12] The idea of the people's subway was so strongly ingrained that foreign visitors often made the mistake of assuming that only poor workers took the subway.[13]

The subways were dirty and unattractive. Station benches that had once been painted bright yellow were now worn, chipped, and greasy, while the platforms were often covered with peanut shells, banana skins, candy wrappers, and old newspapers. The trains were old and drab. During the depression, the subways were flooded with homeless people who slept or panhandled there.[14]

But although the subways were uncomfortable, they were not dangerous. The entire city was much safer in the 1920s and 1930s, and the subways were an extension of that circumstance. A few vicious crimes did take place in the subways, of course. Perhaps the most notorious incident of the interwar years occurred shortly after 2:00 A.M. on May 14, 1936, when a pair of robbers attacked Edgar L. Eckert, a fifty-four-year-old assistant treasurer of the Rogers Peet clothing store, in a bathroom of the Forty-second Street station of the IND's Eighth Avenue line. The robbers strangled Eckert to death and then stole his wallet, gold watch and chain, and gold medallion. Eckert's murder prompted the closing of bathrooms in IND local stations and the assignment of guards to bathrooms in the IRT, IND, and BMT express stations. But violent crimes remained extremely uncommon until the 1960s, and New Yorkers did not fear being killed, robbed, or assaulted on the subways in the 1920s and 1930s. The subways were safe, and they were perceived to be safe. In the lexicon of the New York City subway, the term "safety" referred to train accidents and collisions, not to crime.[15]

Thanks to the nickel fare, New Yorkers could ride the subways and enjoy the city's marvels. But the five-cent fare spelled disaster for the Interborough Rapid Transit Company and the Brooklyn-Manhattan Transit Corporation.

The High Cost of the Nickel Fare

Even after Mayor John F. Hylan had turned the nickel fare into a sacred cause during his 1921 reelection campaign, the Interborough and the Brooklyn-Manhattan continued to try to raise the rate of fare. But IRT and BMT executives recognized that Hylan's success had closed most governmental and political channels; neither elected officials nor transit administrators wanted to invite a public whipping by openly favoring a fare hike.[16]

The Interborough therefore turned to the branch of government that was most capable of resisting popular pressure: the courts. Beginning with New York State courts and then moving to the more conservative federal courts, attorneys for the IRT argued that the nickel fare was confiscatory and asked that the Transit Commission be enjoined to set the fare at seven cents. On May 2, 1928, a three-judge panel of the U.S. District Court for the Southern District of New York ruled in favor of the Interborough. The Transit Commission and the City of New York appealed this judgment to the Supreme Court of the United States. On April 8, 1929, in *Gilchrist v. Interborough Rapid Transit Company,* the Supreme Court reversed the lower court. Writing the majority opinion of this six-to-three decision, Justice James C. McReynolds rejected the Interborough's contention that the existing rate was confiscatory. He also concluded that the company had failed to exhaust its opportunities to resolve the issue in the state courts before bringing suit in federal court. According to McReynolds' strict constructionism, the proper forum for settling this complex question was the state legislature or the state bench. Given New York's political realities, McReynolds' ruling killed the Interborough's last hope of raising the fare.[17]

The nickel fare produced important social benefits by giving the poor access to rapid transit, allowing working-class residents to move to the less crowded neighborhoods in the outlying areas, and emptying the streets of pedestrian and vehicular traffic. The advantages of the low fare were so significant that fares could have been eliminated altogether and the subway made free of charge. But if the fare was to remain low for reasons of social policy, then the IRT and BMT needed to receive direct public subsidies to compensate for lost passenger revenue. In the wake of the Hearst newspapers' attacks on the dual subway contracts' profit-sharing plan and of Mayor Hylan's diatribes against the "transit ring," however, there was no possibility of approving public subsidies. No sane politician would have proposed giving state or municipal funds to the despised subway companies.

This was shortsighted; subsidies would have allowed poor riders access to the system while making the subways profitable. But if the decision had to come down to a choice between keeping the fare at a nickel or raising it, then the cost of a subway ticket should have been increased. Notwithstanding the social benefits of the low fare, its longterm political consequences were ruinous. The financial emergency created by the nickel fare revealed the presence of strains in the existing political arrangements regarding the relationship between the subway companies and government and the legitimacy of public investment in rapid transit. These political problems were never adequately addressed. The nickel fare thus contributed to—and became the abiding symbol of—a political culture that failed to sustain rapid transit over the long run.

In the absence of subsidies, the Interborough and the Brooklyn-Manhattan remained dependent on the fare box. And because the value of the 1904 nickel never rose above 2.68 cents in real dollars during the 1920s, the IRT and BMT continued to experience severe financial trouble.[18] Gone forever were the prewar boom years. Both companies had to struggle to make interest payments on their capital debt. The IRT did not declare any stock dividends during the 1920s, and it reported operating deficits in fiscal years 1923 and 1925.[19]

To survive this tough economic climate, the Interborough and the Brooklyn-Manhattan imposed rigorous cost-cutting measures. Their main target was the labor force. In July 1921 the Interborough Rapid Transit Company slashed workers' wages by 10 percent in order to save $2.6 million in annual operating expenses.[20] The IRT and BMT also abolished thousands of jobs by introducing labor-saving machinery. In a 1920 letter to the Rothschilds, Interborough Chairman August Belmont explained that the fiscal crisis had spurred the IRT to develop mechanical innovations:

> We are gradually installing certain devices, both for the collection of fares and for the operation of the trains, which will mean large savings and keep our corporate life going until something is done [to approve a general plan for the rehabilitation of transit].[21]

One such device was the automatic turnstile. Before the 1920s, people obtained admission to the subways the same way they did to movie houses. A passenger bought a paper ticket from the clerk in the wooden ticket booth and gave it to the ticket chopper at the gate, who cut it in half and admitted the passenger to the platform. The IRT simplified this process in the early 1920s by adopting automatic turnstiles that accepted coins. The result was that the IRT and BMT (which soon adopted the new turnstiles, too) discharged hundreds of ticket choppers. A second machine the Interborough perfected to reduce labor costs was the automatic door control. Before the 1920s, a crew of six employees—a motorman and five guards—was needed to run a ten-car subway train. The reason for this manning level was that the train doors had to be opened and closed mechanically, and each guard could handle the doors of only two cars. By designing automatic door controls that employed pneumatic engines and electrical wiring, the Interborough pared the size of the crew and decreased the number of guards.

By the mid-1930s the standard crew consisted of a motorman who drove the train and a conductor who controlled all the doors. The Brooklyn-Manhattan Transit Corporation eliminated two-thirds of its 4,000 guards between 1919 and 1939, while the Interborough Rapid Transit Company thinned its guard force from 4,000 to 1,519. Altogether, the IRT reduced the number of its employees per million car miles from 85 in 1918 to 45 in 1928. The irony of the people's subway was that rapid transit workers bore the brunt of the economic impact of the nickel fare.[22]

The nickel fare led to disinvestment in underground rapid transit. To reduce capital costs, the IRT and BMT deferred the purchase of new rolling stock and kept older, outdated trains in the fleet longer. To reduce operating costs, the subway companies increased the headway between trains and scheduled fewer runs, cut back on train repairs, and limited the cleaning of platforms, toilets, and cars. The deterioration of subway service was particularly noteworthy on the Interborough, which, except for the problem of overcrowding, had prided itself on maintaining high ridership standards before the war.[23] In 1927, IRT President Frank Hedley expressed the company's new approach by saying, "I saw a car with clean windows today, and when I got back to the office I raised hell to find out who cleaned those windows and spent all that money."[24] A few riders saw the connection between the five-cent fare and the poor quality of passenger service. One English visitor, E. Stewart Fay, gave his observations:

> Perhaps it is because of the fare war that the subways can find no money to spend on anything but transport. . . . Nothing is done to relieve the stark ugliness of the stations or the trains, and little is done to ventilate them, and it is impossible for a Londoner to gain a first impression other than highly unfavourable to New York's rapid transit.[25]

Although this retrenchment enabled the BMT and IRT to become profitable by the mid-1920s, the great depression devastated the two companies. From 1928 to 1937 the BMT's rapid transit ridership dropped 12 percent, and the IRT's patronage fell 19 percent.[26] During that period the BMT's net annual income shrank from about $6.6 million to about $4.6 million. The IRT was hit much harder. In addition to its severe traffic losses, the Interborough was also burdened by the onerous terms of its 1903 lease of the Manhattan Railway Company. As modified in 1922, this lease required that the Interborough's holding company pay the Manhattan bondholders a fixed guaranteed rental of 5 percent on the elevateds' valuation of $60 millon. This lease siphoned income from the IRT's relatively healthy subways to the owners of its obsolete, deficit-plagued els. Between 1930 and 1939 the total payments made to Manhattan bondholders amounted to $27.8 million, almost equal to the $30 million in profits that the IRT subways cleared during this same period. In August 1932 the Interborough management took the company into receivership, largely in an attempt to disavow the Manhattan lease and concentrate on the more successful subway.[27]

The great depression endangered the survival of the Interborough Rapid Transit Company and the Brooklyn-Manhattan Transit Corporation as independent companies and threatened the integrity of the rapid transit system. The bankruptcy of the Interborough meant, for instance, that the IRT subway and the Manhattan els might be severed and operated as two distinct networks, at higher fares and with no transfers. The reorganization of the subways thus became a central concern of New York City Mayor Fiorello H. La-Guardia.[28]

Fiorello H. LaGuardia and Unification

When Fiorello H. LaGuardia took the oath of office as New York City's ninety-ninth mayor shortly after midnight on January 1, 1934, his chief supporter, Judge Samuel Seabury, happily exclaimed, "Now we have a mayor of New York!"[29]

LaGuardia was born on December 11, 1882, in New York's Greenwich Village; he was the son of middle-class immigrants who had recently arrived from Italy. His father, Achille, soon enlisted as a bandmaster in the U.S. Army's Eleventh Infantry Regiment, and Fiorello grew up on a series of western military posts. Following Achille's discharge from the army in 1898, the family sailed across the Atlantic and settled in Trieste, Austro-Hungary.

The Little Flower, as he was called, returned to New York City in 1906. After graduating from the evening division of New York University's Law School in 1910, LaGuardia opened a law practice at 15 William Street in lower Manhattan, handling deportation, estate, tenant, and other penny-ante cases for a largely Italian immigrant clientele. Eager to enter politics, LaGuardia became active in lower Manhattan's Madison Republican Club. He avoided the dominant Democratic party partly because he despised the corruption-riddled Tammany Hall and partly because he realized that an Italian-American probably would not get very far in that Irish-led machine.[30]

In 1916, LaGuardia won an uphill battle for Congress in lower Manhattan's heavily Democratic 14th Congressional District. After serving as president of the New York Board of Aldermen in 1920–21, he was reelected to Congress in 1922 from East Harlem's Italian and Jewish 20th C.D. Plump, five feet two inches tall, with a shock of black hair that fell across his forehead, and usually wearing an ill-fitting suit and a fedora, the flamboyant LaGuardia acquired a national reputation as a fighting progressive in the 1920s. He assailed Secretary of the Treasury Andrew Mellon's tax policies as a give-away to the wealthy and as a threat to economic prosperity, railed against Prohibition as an infringement on personal freedom, and denounced the restriction of immigration as racist. When LaGuardia ran for reelection to Congress in 1932, however, he was beaten by his Democratic opponent, who rode the Roosevelt tidal wave to victory.

This stunning defeat convinced many observers that fifty-year-old LaGuardia's political career was over. But a corruption scandal that rocked the city soon gave him new life. In August 1930, Judge Samuel Seabury was named to investigate New York City's magistrate courts. Over the course of the next two years Seabury conducted a series of wide-ranging hearings that revealed the Tammany-controlled municipal government had sold judicial appointments, zoning variances, and official licenses under the table; city leaders hobnobbed with mobsters and gamblers; the district attorney's office quashed indictments for a price; and the police department protected numbers games, racetrack fixes, and labor racketeers and framed innocent women on phony prostitution charges. These sensational disclosures forced Mayor Jimmy Walker to resign from office in September 1932. Although Walker was succeeded first by the Democrat Joseph V. McKee and then by the Democrat John P. O'Brien, the scandal wounded Tammany Hall. Backed enthusiastically by Seabury, Fiorello H. LaGuardia secured the mayoral nomination of the reformist Fusion party. In November 1933, LaGuardia was elected mayor. It was hardly a landslide victory, however; the Little Flower failed to receive a majority of the ballots and would have lost the race if his two rival Democratic candidates had not split the opposition vote. LaGuardia tallied 868,522 votes, compared to 609,053 for McKee and 586,672 for O'Brien. The result was that LaGuardia would be operating from a weak position and, despite the scandals, Tammany Hall would remain potent.[31]

LaGuardia became chief executive at perhaps the bleakest moment in New York City history. The great depression had ravaged the city. Tens of thousands had lost their jobs and could not pay rent or buy food. Bread lines, soup kitchens, bank closings, and shantytowns made New York a picture of misery. True to form, LaGuardia leaped into action. After sixteen years of Tammany corruption and lethargy, New York finally had a dynamic mayor who was determined to bring a new ethic of honesty, efficiency, and economy to government and to revive public confidence in the city's future. A tireless, fiercely combative man who insisted on tackling every detail and hated to delegate authority, LaGuardia personified "'hands-on'" management. "It seemed as though the town had been invaded by an army of small plump men in big hats," historian William Manners later wrote. "He was everywhere."[32] Fiorello H. LaGuardia brought about major changes in local government and in the city's landscape. He probably deserves his reputation as the best mayor in New York history.[33]

One of the few policy areas where LaGuardia kept his distance, however, was rapid transit. "I don't think LaGuardia paid too much attention to this [rapid transit] situation," recalled Goodhue Livingston, the mayor's executive secretary.[34] According to Livingston, LaGuardia viewed transit as a source of potential problems that could only hurt him politically. The biggest problem, of course, was the possibility of a fare hike. LaGuardia was convinced that the five-cent fare was a boon for the ordinary guy who had to dig deep into his wallet to pay his landlord, put food on the table, and clothe his family. As mayor, LaGuardia's commitment to the nickel fare was reinforced by his realization that an increase would provoke a violent public reaction that would hurt him politically. LaGuardia himself had already exploited this issue when, as president of the Board of Aldermen in 1920–21, he had made demagogic attacks on his enemies in the Transit Commission and the Republican hierarchy for supposedly planning to raise the subway fare.[35] LaGuardia's position on mass transit was very close to Mayor Hylan's urban populist stance. As mayor, LaGuardia wanted to avoid having to raise the turnstile cost because he realized only too well that a fare hike was —in Livingston's words—"sacrilegious."[36]

Another example of LaGuardia's negative perception of rapid transit as a source of problems involved his decision to keep John H. Delaney as chairman of the Board of Transportation. A onetime Brooklyn printer and an official of a typographer's local, Delaney rose to an important position in the Tammany machine through his friendship with Charles F. Murphy, the legendary Tammany boss. Delaney, who obtained public office through his political connections, proved to be a competent administrator. Appointed chairman of the Board of Transportation by Mayor Hylan in 1924, the hard-nosed Delaney mastered his agency and became known as "Mr. Transit." Although LaGuardia ousted most Tammany stalwarts from office, he kept Delaney. This was a wise decision. As master builder Robert Moses said admiringly, Delaney proved to be "a smart cooky."[37] He was personally loyal to LaGuardia and implemented his policies. Moreover, Delaney oversaw the construction of the IND and ensured that no political scandals occurred that might embarrass LaGuardia.[38]

In his preoccupation with the subways and els as a source of difficulty, Mayor LaGuardia missed rapid transit's positive contributions to New York City. He simply did not give a lot of thought to rapid transit, except when he had to. Unlike fascinating new technologies such as the airplane (which he loved), LaGuardia viewed the subways as old and thus uninteresting. He certainly did not value the role the els and subways played in stimulating New York's tremendous physical expansion since the 1870s or their potential as a spearhead for further urban development.[39] Although not as automobile-minded as Robert Moses, LaGuardia strongly backed the highways, bridges, and tunnels that Moses was constructing with federal funds in the 1930s. To LaGuardia the automobile, not rail rapid transit, embodied New York's best hope for the future.[40]

No transit problem that confronted LaGuardia was more important than the restructuring of the private subway companies. A number of plans had been proposed since the State of New York first authorized the reorganization of transit in 1921, but the great depression gave new urgency to the matter.[41] The IRT's and BMT's desperate financial woes raised the possibility that the rapid transit systems would disintegrate.

Transit reorganization was part of a larger problem—the City's severe fiscal crisis. According to historian Thomas Kessner, by 1933 New York City was "in the throes of its most extreme financial crisis in modern history."[42] At the very time that the city had to shoulder the tremendous burden of providing relief for the growing numbers of unemployed, its resources were shrinking. The Hylan and Walker administrations had spent lavishly on capital projects in the 1920s, leaving the city with a mountain of debt. The city was able to cover this debt due to steady increases in its tax receipts during the booming 1920s, but its revenues contracted when the depression began. In 1932 the total value of tax delinquencies climbed to $79 million and accounted for 15 percent of all taxes owed. Unable to fund its expenditures out of current income and grappling with huge debt obligations and skyrocketing relief bills, New York City resorted to the desperate expedient of borrowing against projected receipts. By 1932 the city had amassed a debt that was nearly equal to that of the forty-eight states combined.[43]

The subways had contributed mightily to this financial emergency. Part of the problem stemmed from the dual contracts' profit-sharing arrangement. By the late 1920s the city was receiving only $2.65 million per year from the Interborough and the Brooklyn-Manhattan on its investment of $386,050,000. Without the revenues anticipated from the dual system, the city had to pay its debt charges from the tax budget. These annual charges amounted to $15 million in 1931. This was money that the hard-pressed city could not afford.[44]

The City of New York was also squeezed by the Independent Subway System, which was proving to be extremely expensive to build. The IND cost an average of $9 million per route mile, 125 percent higher than the earlier IRT and BMT subways. By June 1939 the IND's total construction expenditures had reached $766,844,000. The municipality also had to underwrite the operation of the Independent. Shortly before the first IND line entered service in 1932, the Board of Transportation realized it would incur huge annual operating deficits. Deciding to abandon Mayor Hylan's old dream of municipal operation, the board drew up a contract, known as Contract No. 5, for the private operation of the new system. But private businessmen could read a balance sheet, too. When the city sought bidders for this contract later in 1932, the only formal response came from the Oakdale Contracting Company, which insisted on receiving subsidies for the maintenance and operation of the new subway.[45] The Board of Estimate rejected Oakdale's proposal and reluctantly approved the IND's opening as a municipally owned and run facility. As feared, the IND became a big drain on the municipal treasury. Every IND trip cost fourteen cents in this era of a nickel fare, so the city lost nine cents per ride.[46] As a result of the grave economic climate of the 1930s, public operation of the IND was no longer viewed as the bold experiment in governmental innovation that Mayor Hylan had proclaimed ten years earlier. Rather, municipal operation was now a temporary expedient that had to be endured until a general solution to transit's financial ills could be found.[47]

To Fiorello H. LaGuardia, reorganization of the subways was primarily a way to restore the city's finances. He contended that unification of rapid transit under public control would save millions of dollars in capital costs by eliminating the dual contract preferentials and allowing interest costs and sinking fund charges to be paid from subway revenues rather than out of taxes. The mayor also thought that unification would permit the city to refinance corporate securities at much lower interest payments. This refinancing would not only reduce the city's interest charges and retire its transit debt more quickly but would also free capital that could be used for the construction of new subways, schools, parks, hospitals, and housing. In addition, LaGuardia expected that the combination of the three separate systems—the IRT, BMT, and IND—into a single, integrated network would produce scale-operating economies that would save millions of dollars more. He put great faith in the efficiencies that could be achieved by combining the three subways under one management.[48]

There were three central elements in LaGuardia's conception of unification. The first element was his unshakable commitment to the five-cent fare; there was never any doubt that the subways would continue to be operated on the basis of the nickel fare. The second component involved the kind of government body that would take over the subways and els. LaGuardia was determined that the city would run the subways and els itself, a position that represented a significant departure from prior reorganization proposals. For instance, in 1931 the Transit Commission had proposed a reorganization plan that envisioned power sharing between the public and private spheres. Under the terms of this plan, the City of New York would acquire the rapid transit properties of the Interborough Rapid Transit Company and the Brooklyn-Manhattan Transit Corporation for $474.5 million and then lease them to a quasi-public corporation known as the Board of Transit Control that would operate the railways. This board would consist of nineteen members, including ten who were to be appointed by the mayor, five who were to be named by the private transit investors, and several public officials serving *ex officio*.[49] In contrast to this TC proposal, LaGuardia wanted the municipal government to operate the railways itself, but there were pitfalls to his approach. At the very moment New York City was losing its financial self-sufficiency and President Franklin D. Roosevelt's national government was assuming more responsibility for urban functions such as relief, public housing, and highway construction, LaGuardia proposed to bring a rapid transit system that was losing money under municipal control.

There was little likelihood of obtaining federal money for the subways. Due to their evolution as profit-making businesses confined to particular cities, mass transit continued to be perceived in local rather than in regional or national terms. In May 1934, however, the LaGuardia administration did win a grant and loan of $23,160,000 from the Public Works Administration for the building of the Independent Subway System. This marked the first time the federal government ever aided New York transit, and it was clearly an exception. To LaGuardia as well as nearly all transit experts and government leaders in the 1930s, the idea of federal support for local railway operation was unthinkable.[50]

The third component in Mayor LaGuardia's view of reorganization was his conviction that the subways should be self-supporting. Anticipating that unification would result in great operating economies, LaGuardia expected them to be self-sustaining even with the retention of the nickel fare. Largely because of its development as a profit-making business, transit was perceived differently from schools, fire and police departments, and sanitation; these municipal services did not have to be self-sustaining. Public subsidies for transit would smack, at best, of financial waste and irresponsibility; at worst, of socialism. Following the controversies that had arisen in the late nineteenth and early twentieth centuries over the rise of urban railway monopolies and the poor quality of passenger service, governments passed stringent regulations aimed at controlling urban transit companies. The conflict-ridden relationship between government and management was so powerful in contemporary minds that even transferring operation from the private to the public sector could not erase the underlying image of transit as a profit-making enterprise. Long after private operation of rapid transit ended, the subways continued to be saddled with this ideology of business management.

LaGuardia acted within the context of transit's political culture and, except for trying to bring the privately owned subways under public control, never directly challenged its main tenets. He approved of the antagonistic relationship between business and government, endorsed the public's demand for the nickel fare, and conceived of rapid transit in local rather than regional or national terms. Above all, LaGuardia accepted the traditional definition of the subways as a business and was not prepared to articulate a new understanding of New York's transit system as a vital municipal enterprise that warranted public subsidies. Embroiled in the worst budget crisis in the city's recent history, determined to bring economy to government, and feeling no enthusiasm for rail rapid transit, LaGuardia sought to make the subways self-supporting and drive them away from the public trough.[51]

From Private to Public

Under the provisions of the 1921 act that sanctioned the restructuring of New York's mass transit companies, only the Transit Commission had the power to reorganize the subway companies. This statutory restriction displeased Mayor LaGuardia. He believed the city, not the state, should determine the fate of the subways. He also disliked the Transit Commission because it was a Tammany stronghold. Early in his first term LaGuardia tried to transfer the power over unification from the Transit Commission to the municipal government, only to have the Democratic legislature bury the proposal. Then he recommended that his own aides be authorized to speak for the Transit Commission, only to have the TC torpedo that idea. Unable to accomplish anything through regular channels, the mayor brashly seized the initiative by naming his own representatives and instructing them to proceed as if the TC did not exist.[52]

To conduct the negotiations, LaGuardia chose two leading

members of his administration, Judge Samuel Seabury and Adolf A. Berle, Jr. Seabury, of course, was best known for investigating Jimmy Walker's corrupt administration and for engineering LaGuardia's nomination. A descendant of eight generations of New Englanders, the namesake of the first Episcopal bishop in America, and a respected member of the Anglo-Saxon elite, sixty-year-old Seabury was a square-jawed, Olympian figure who was aloof and self-righteous.[53]

Born in 1895 and nicknamed the "boy prodigy" for his intellectual brilliance, Adolf A. Berle, Jr., entered Harvard College at age fourteen, received his B.A. at eighteen, and graduated *cum laude* from Harvard Law School at twenty-one. In 1932, Berle, by then a law professor at Columbia University, and economist Gardiner C. Means wrote *The Modern Corporation and Private Property*. Acclaimed as one of the twentieth century's most important economic treatises, Berle and Means' book contended that the modern corporation had become independent of its stockholders and that management competed on a nearly equal basis with the modern nation-state. Their book advocated the enactment of strong economic regulations to control these runaway corporations. Published in the depths of the great depression and offering an explanation for the nation's economic miseries, *The Modern Corporation and Private Property* came to the attention of New York State Governor Franklin D. Roosevelt, who made Berle a member of his brain trust. After his election as president in 1933, Roosevelt appointed Berle counsel of the Reconstruction Finance Corporation. Abrasive, edgy, quick to make enemies, Berle was one of the brightest New Dealers. The following year, while retaining his New Deal job, Berle joined LaGuardia's reform administration as city chamberlain, assigned to mastermind the restructuring of municipal finances. Functioning as a "prime minister without portfolio," Berle took care of everything from settling elevator strikes to negotiating with bankers to restore the city's credit and acquiring more federal aid.[54] Part of Berle's efforts to end the city budget crunch involved his attempt to unify rapid transit.[55]

LaGuardia, a bad-tempered boss who ranted, raved, and screamed at his subordinates, never bullied Berle or Seabury. To the contrary, he treated these well-born gentlemen with respect and deference; indeed, he never even called Judge Seabury by his first name. Having to reorganize the subways in order to solve the city's budget squeeze, LaGuardia needed a settlement in spite of his own political weakness as a minority mayor and the city's lack of statutory authority over reorganization. So he picked two men of great stature, delegated power to them, and waited for developments.[56]

Seabury and Berle, who began negotiating with the BMT and the IRT early in 1934, confronted a task of enormous proportions. Transferring the properties of these two large industrial corporations to public ownership was remarkably complicated; simply to list the properties that were to be included in these purchases took dozens of pages in memoranda and documents. In addition to the sheer size of the acquisition, Seabury and Berle had to contend with the intricacies of corporate ownership. Not only did they have to reach terms with the IRT's and BMT's executives, but both companies' security holders had to approve the agreements; owners of each class of stocks and bonds had an opportunity to accept or reject the deal. Seabury and Berle's discussions were further complicated by the Interborough's bankruptcy, since the Federal District Court that was overseeing its receivership had to consent to the transactions.

The IRT's and BMT's managers and investors generally favored unification, however. As the great depression deepened and the prices of their securities plummeted, investors saw unification as a means of cashing in their stocks and bonds before they lost even more value. Moreover, company officials and investors knew that if unification did not succeed, the city could always take over the subways by invoking the recapture clauses of the dual contracts. Even though the IRT and BMT would be compensated for the recaptured lines, the city would take the heart of their systems and leave the companies with skeletal networks that would be difficult to operate independently. To the IRT and BMT, unification through negotiation and purchase was preferable to recapture. The companies were determined, of course, to drive a good deal: If the city took over the subways and brought private transit ownership to an end, they wanted to receive the highest purchase price possible; if private investors retained a property interest in rapid transit, then they wanted to abolish the nickel fare and ensure the subways' profitability.[57]

Seabury and Berle soon reached preliminary agreements with the two companies. They signed memoranda of understanding with the BMT on February 19, 1935, and with the IRT on November 1, 1935. Although these memoranda constituted rough working documents rather than definitive purchase plans, they marked the first time that representatives of the city and companies had ever achieved a formal accord that set a price. Under the terms of these memoranda, the City of New York would pay a total of $430,751,000 for the IRT's subway and elevated divisions, the BMT's rapid transit lines, and the BMT's Williamsburg power plant. The city would lease the BMT and IRT properties, plus the IND subway, for seventy-five years to an independent government corporation, the Board of Transit Control, which would operate the subways and els as a unified system with a five-cent fare. It would be governed by a board of fifteen directors, eleven appointed by the mayor with the approval of the Board of Estimate and four selected by the mayor upon nomination of the company security holders.

Seabury and Berle's proposals envisioned a greater measure of public control over rapid transit than had any prior reorganization plan. But they failed to achieve Mayor LaGuardia's goal of municipal ownership and operation. Their Board of Transit Control was an autonomous government corporation rather than a line agency of the municipality, like the Board of Transportation, and it would give transit investors a continuing voice in transit decisions through their membership on the board of directors. Seabury and Berle devised the Board of Transit Control mechanism to circumvent the state constitution's imposition on New York City of a debt limit of 10 percent on the assessed valuation of property. Due to the extravagant public works spending of the 1920s and the reduction of real estate taxes during the depression, the City of New York lacked the borrowing capacity to finance unification. As a semi-independent corporation free of the city's fiscal restraints, the Board could issue its own bonds. According to Seabury and Berle's agreements, the Board could raise three-quarters of the $430 million purchase price through its bond issues, and the city would have to cover only 25 percent of the price, in the form of cash or city securities.[58]

After signing the preliminary agreements, Seabury, Berle, and the company representatives prepared a final proposal called the

Seabury-Berle plan, which was similar to the memoranda of understanding except that the purchase price was increased from $431 million to $436 million. On June 22, 1936, Seabury and Berle submitted their plan to the Transit Commission for approval, as required by law.

Ten months later, on May 7, 1937, the TC rejected the Seabury-Berle plan in a blistering assault that condemned the LaGuardia administration for trying to bail out the IRT and BMT. The TC claimed that the Seabury-Berle plan would handcuff the city by giving the bankers control of its transit system for seventy-five years. It also contended that the purchase price was excessive and that the plan would lead to either higher taxes or a fare hike.[59] TC special counsel James J. Curtin accused the LaGuardia administration of mounting "the most dangerous and insidious attack on the five-cent fare that has ever been launched in transit's turbulent history."[60]

These Transit Commission charges were baseless.[61] Part of the TC's antagonism stemmed from the traditional state-city rivalry that pitted the State Transit Commission against the municipal government. The main reason, however, was that the Tammany Democrats who dominated the Transit Commission were trying to use unification as a weapon to unseat Fiorello H. LaGuardia in the 1937 mayoral election. This report amounted to an opening salvo in the campaign. Three days later LaGuardia went on the radio and blasted the TC. He vigorously defended the Seabury-Berle plan and insisted that rapid transit unification would preserve the nickel fare. He berated the TC for being the creature of "political bosses" and for being a "sham," a "fraud" and "the latest word in political duplicity."[62]

On November 2, 1937, LaGuardia won a landslide victory, capturing 1,344,016 votes to his Democratic opponent's 889,591 votes. He became the first reform mayor to succeed himself in New York City history. His allies in the Fusion movement gained control of the Board of Estimate, too.[63] Although LaGuardia's reelection did not turn on transit issues, it had a major impact on unification. His drubbing of Tammany took the fight out of the Transit Commission and put it on the defensive. After 1937 the TC no longer posed an obstacle to unification.[64]

On May 14, 1938, Mayor LaGuardia announced that the city

and the subway companies would resume their talks. A new set of negotiators would conduct these sessions. After LaGuardia's reelection, Adolf A. Berle returned to Washington, D.C., to become assistant secretary of state, while Samuel Seabury retired to private life. Their replacements were John H. Delaney, chairman of the Board of Transportation; Newbold Morris, a reform lawyer who had been elected City Council president on the Fusion ticket in November 1937; and Joseph D. McGoldrick, a former professor of government at Columbia University and the Fusion party's victorious candidate for comptroller in the previous election. As mavericks who commanded independent reputations that exceeded the bounds of any official position, Samuel Seabury and Adolf A. Berle had been well suited to asserting the municipality's case at a time when LaGuardia was politically weak. The new team of Delaney, Morris, and McGoldrick reflected the stronger position that LaGuardia enjoyed during his second term. They held public office, knew how to cooperate with other people, and liked to work behind the scenes.[65]

Along with the 1937 election, the city's negotiating posture was also enhanced by the so-called Roosevelt depression of 1937. This economic catastrophe sent transit securities into a nosedive. Between June 22, 1936, when Seabury and Berle submitted their unification proposal to the Transit Commission, and May 14, 1938, when Mayor LaGuardia disclosed that the unification talks would begin again, the IRT's 5 percent bonds dropped from $94\frac{3}{4}$ to $57\frac{3}{4}$, the IRT common stock tumbled from $14\frac{7}{8}$ to $5\frac{1}{8}$, and the BMT preferred stock plunged from 102 to 29. This steep decline enabled the city to drive a tougher bargain by making investors eager to sell before their securities slipped even further.[66]

Consequently, the new round of unification conferences went more smoothly. On July 1, 1939, Mayor Fiorello H. LaGuardia and BMT president William S. Menden signed a contract to sell the company's properties for $175 million. Four months later, on November 1, 1939, LaGuardia contracted to acquire the IRT for $151 million. These pacts were superior to the Seabury-Berle plan. The 1939 contracts set a total purchase price of $326,248,000, a third less than Seabury-Berle's figure of $436 million. These savings were doubly impressive because the 1939 agreements included more transit properties. The 1936 plan had been confined to the BMT and IRT subways and els, but the decrease of transit securities had permitted the Brooklyn-Manhattan's extensive street railways and buses to be added to the 1939 package. Because these Brooklyn surface lines were valued at $27 million, the 1939 plan represented a savings of nearly $137 million. Moreover, these new contracts dispensed with Seabury-Berle's mechanism of the Board of Transit Control and provided for the Board of Transportation to run the subways, els, street railways, and buses. For the first time a unification plan authorized municipal ownership and operation of transit. Although some critics still complained that the purchase price was excessive and that the plan constituted a sellout to the IRT and BMT investors, it was a good deal for the city.[67]

But the city's debt ceiling was one obstacle that threatened to impede unification. Although the 1939 plan fixed a better price than Seabury-Berle had and although the city was in stronger financial shape than before, the municipal government nonetheless lacked the credit needed to fund the acquisition. Fortunately, a convention was meeting in Albany to draw up a new state constitution. LaGuardia's assistants drafted an amendment to exclude the $315 million required for unification from the city's debt limit. State Transit Commissioner M. Malcolm Fertig cooperated by shepherding the proposed amendment through the convention, which adopted it in August 1939 and put it on the ballot in the November 1939 general election. Backed by business, labor, and civic groups, Mayor LaGuardia campaigned for the ratification of the amendment. Upstate voters rejected the measure, but the city's electorate approved it by a margin of better than two to one, enough to ensure its passage statewide. The ratification of this amendment allowed the municipality to finance the purchase by issuing 3 percent, forty-year city securities.[68]

The next step was for the BMT and IRT investors to signify their approval of the plan by depositing their securities. The consent of 76 percent of the IRT and Manhattan senior lien bonds and 66 percent of the Manhattan capital stock was required before the Interborough contract became operative. The IRT deposits proceeded on course, and the Transit Commission declared the IRT plan operative on November 22, 1939. Four months later Federal Judge Robert

M. Patterson approved the purchase of the IRT and ended the company's seven-year receivership. The BMT contract would become operative when two-thirds of the company's voting stockholders and 90 percent of the holders of its preferred stock and various classes of bonds made their deposits. Although most investors deposited their securities, the owners of several bond classes demanded better terms and refused to settle. LaGuardia tried to intimidate the holdouts by threatening to recapture the BMT and leave the company with a small system that would be almost worthless. "We will dismember the BMT so you won't be able to recognize it," LaGuardia blustered. Nobody believed him, however. LaGuardia finally had to sweeten the deal by raising the price of some bonds. The largest holdout, Prudential Insurance Company, received par for its $3.1 million of BMT collateral bonds. More deposits were made, and the BMT plan went into effect in March 1940.[69]

At 11:55 P.M. on May 31, 1940, the last BMT train under company management rolled out of the Fifty-seventh Street–Seventh Avenue station. This special train bore such prominent guests as Mayor Fiorello H. LaGuardia, his wife Marie, BMT president William S. Menden, and Board of Transportation chairman John H. Delaney. Four minutes later the special entered the Times Square station, where a crowd of five hundred people was waiting on the platform. They had come to witness a ceremony commemorating the transfer of the BMT properties from private to public ownership. At midnight on June 1, 1940, the hour when the title passed to the city, the train's horn gave three shrill blasts. President Menden formally turned over his company's properties to Mayor LaGuardia who in turn gave them to Chairman Delaney. Delaney then appointed LaGuardia as Motorman No. 1 of the New York City Transit System and presented him with a badge and a special set of silver motorman's tools encased in a shiny wooden box. Mayor LaGuardia—who loved nothing as much as a good show—clicked his heels and saluted smartly. "I am glad to be working for you," the mayor said, "and I hereby report for duty."[70] LaGuardia then donned a motorman's cap and jacket and posed for photographers in the train's cab. Although the mayor could not resist fiddling with his brake handle and control lever and blowing the horn a few times, he soon surrendered the instruments to a regular BMT motorman, John Donnellan. As Donnellan eased the special out of the station and resumed its run to Canal Street, the first car bore a huge banner that proclaimed FIRST TRAIN, B.M.T. DIVISION, NEW YORK CITY TRANSIT SYSTEM. A week and a half later, the city took title to the IRT's properties. New Yorkers continued to refer to the two divisions by the familiar names BMT and IRT, but the Interborough and Brooklyn-Manhattan companies were no more. A part of New York City history was gone forever.[71]

Unification was a massive enterprise. It constituted the largest railroad merger in U.S. history and the largest financial transaction undertaken by the City of New York. To sign and transfer the checks and legal documents required for the BMT acquisition took an entire morning. The New York City Transit System (NYCTS) was an immense enterprise, with nearly thirty-five thousand employees (including twenty-six thousand transferred from the private subway companies) and an annual payroll of $60 million. The NYCTS was the biggest transit system in the world, a vast network of 760 track miles of subways and els, 435 miles of street railways, and 80 miles of bus lines. It was also the most heavily used system in the world; during its first full year of operation the NYCTS carried a mind-boggling total of 2.3 billion passengers, 1.8 billion of whom were on its subway and elevated lines. According to historian Joshua B. Freeman, 71 percent of all New York City residents rode the system in a typical month. On average, every New Yorker took 26.5 trips and spent ten hours riding the system per month.[72]

One way or another the transit system touched almost everyone in the city. Without it, New York could not have existed.

10

THE REVOLT AGAINST POLITICS

"We Made a Good Deal"

Shortly after Mayor Fiorello H. LaGuardia presided over the unification ceremony in the BMT's Times Square station on the first day of June 1940, New Yorkers rediscovered an old problem: passenger overcrowding. Transit patronage had fallen during the great depression, but it soared in the 1940s. The war economy fattened civilian paychecks at the same time that its rationing of gasoline and rubber emptied the highways of mass transit's great competitor, the automobile. From its low point of over 1.7 billion in 1933, New York's subway and elevated ridership increased to over 1.9 billion in 1943 and then peaked at more than 2 billion in 1947.[1]

This surge discomforted many straphangers. In 1944 a *Saturday Evening Post* journalist described rush hour as "Manhattan's daily riot" and complained that other commuters pounded on him like "human pistons" in the "crashing, clashing, grating cacophony of the trains."[2] Writing in the *New York Times Magazine* two years later, Murray Schumach called the underground railway "a mobile torture chamber" where no one had enough elbow room to look at his wristwatch, button his coat, or pull out his handkerchief before coughing or sneezing.[3]

But Mayor LaGuardia knew the packed trains meant the great depression had finally ended. LaGuardia exclaimed in 1943:

> Let me tell you this. Any time we don't have crowding during the rush hour, there'll be a receiver sitting in the mayor's chair and New York will be a ghost town. Why, they talk about the rush hour and the crash and noise. Why, listen, don't you see that's the proof of our life and vitality? Why, why, that is New York City![4]

LaGuardia was also pleased because the heavy traffic enabled the transit system to make a profit. With nearly 2 billion riders passing through the turnstiles every year, Mayor LaGuardia and his lieutenants were confident that unification was an unqualified success. "We made a good deal," former City Council president Newbold Morris proclaimed in 1950. Morris, who had helped negotiate the 1940 unification accords, boasted that the LaGuardia administration left office in December 1945 with an operating surplus of $25 million.[5]

But these rosy traffic and financial figures were misleading; in reality, New York's rapid transit system was in trouble. World War II ended in 1945, and Detroit's factories soon began to churn out automobiles once again. This renewed competition from motor vehicles wrecked the country's urban railways and buses. Between 1945 and 1960 total mass transit ridership in the United States plunged from 23.4 billion to 9.4 billion passengers, while automobile registrations rose from about 25.8 million to about 61.7 million. In this same fifteen-year period, New York City's bus, trolley, subway, and elevated ridership dropped from approximately 2.4 billion to 1.8 billion. Its rapid transit ridership—that is, its elevateds and subways—slipped from 1.9 billion to 1.3 billion.[6]

These ridership losses put great financial strain on the New York City Transit System and undermined the 1940 unification agreement. This decrease suggested that Mayor LaGuardia was unrealistic to expect that the subways could be put on a self-supporting basis with a nickel fare and that the integration of the separate IRT, BMT, and IND systems could produce significant operating economies. "I thought that the whole idea of unification was misplaced," former Transport Workers Union leader Maurice Forge said in a 1987 interview. "It was a patchwork solution to an impossible patchwork problem."[7] Forge questioned LaGuardia's basic assumption that two decrepit, badly neglected private companies on the verge of financial collapse could be successfully combined with a municipal subway that had to be heavily subsidized to stay afloat. According to Forge, unification was flawed because it did not provide the subways with a reliable source of funding. He thought a rational solution would have required the city to raise the fare high enough—to ten or fifteen cents—to operate the subways adequately and to commit more tax revenues to capital improvements. But Forge readily acknowledged that such a rational course of action would have been politically impossible. Because city politicians fretted that an increase would provoke a furious public outcry akin to that of "doing damage to motherhood, apple pie, and the Constitution," he conceded that "nobody wanted to bell the cat."[8]

Once again a fiscal crisis revealed cracks in the political system and created pressure for change. The failure to correct subway finances led a prominent Republican attorney, Paul Windels, to campaign for the abolition of direct municipal operation of rapid transit.

Paul Windels

Paul Windels was born in Williamsburg, Brooklyn, on December 7, 1885. His father, a German-born Methodist minister, caught pneumonia and died on Christmas Eve 1899. To support the family, fourteen-year-old Paul left school to go to work as an office boy for an importing firm in lower Manhattan. A bright and ambitious young man who read widely, Windels won a scholarship to Columbia College in 1904. He received his B.A. four years later. In 1910 he graduated first in his class from Brooklyn Law School and then began to practice with the firm of Emmett, Parish and Emmett at 52 Wall Street. Seven years later he formed his own partnership, Windels & Holtzoff. Tall, erect, and dignified, with frugal habits and a strong work ethic, Windels had firm moral convictions. A progressive Republican who served as a delegate to the GOP national conventions of 1920, 1924, and 1928, he believed that government should be small in scale, efficient, and honest. He was also a skilled political tactician. He managed Fiorello H. LaGuardia's successful campaign for president of the New York City Board of Aldermen in 1919. In the early 1930s, Windels aided reformers Samuel Seabury and Charles C. Burlingham in the battle to rid the Tammany-led city hall of corruption and fiscal recklessness. When LaGuardia entered city hall in 1934, Windels became his corporation counsel. He rooted out corruption from the law department, eliminated its huge case backlog, and implemented managerial improvements that saved the treasury nearly $50 million. He returned to private practice following LaGuardia's reelection in 1937.

In the late 1930s Paul Windels grew apprehensive about the growth of big government and about the growing support for radicalism in the United States, and he began to move in a more conservative direction. In 1940, Windels became chief counsel of the Rapp-Coudert legislative committee's investigation of the perceived communist threat in New York City's public schools and colleges. He helped bring about the dismissal of nearly fifty teachers and clerks for supposedly "subversive activities."[9]

In 1940, Windels started a drive to cut municipal expenditures by founding a civic group, the Committee of Fifteen. Backed by realtors and other businessmen who wanted to reduce property taxes, the Committee of Fifteen protested that New York's bloated real estate levies put local businesses at a competitive disadvantage with firms in other American cities. It claimed that wasteful spending threatened to eliminate thousands of private-sector jobs, destroy the municipal tax base, and halt the city's recovery from the depression. Windels warned that big government endangered "our progress as a city."[10]

The Committee of Fifteen criticized the LaGuardia administration for numerous instances of municipal extravagance, including its decisions to raise civil service wages and pensions, install a new three-platoon system for the fire department, and expand welfare, hospitals, parks, and other services. The committee's primary target, however, was the subway. "The log which was really causing the logjam," Windels said, "was the so-called transit deficit."[11] Even though the New York City Transit System turned an operating surplus every year during World War II, it required a large debt service

—$37 million by 1944—and constituted the single biggest drain on the municipal treasury.[12]

Paul Windels regarded the subways as a business and adamantly opposed public subsidies for operating or capital charges. He blamed the subway mess on the nickel fare. In a 1943 survey of the country's twenty-five largest cities, the Committee of Fifteen found that only New York retained the straight nickel fare. These civic leaders and businessmen thought that a rate increase was long overdue. In 1944, Windels formed a new organization, the Citizens' Transit Committee, to lobby for a higher fare.[13]

Like the Committee of Fifteen, the Citizens' Transit Committee was dominated by realtors and businessmen. The participation of realtors was particularly noteworthy. Developers had championed the building of the elevated railways, the 1904 IRT underground railway, and the dual contract subway in order to stimulate residential construction in the outlying areas. But their enthusiasm for rapid transit slackened in the 1920s for two reasons. First, Mayor John F. Hylan's decision to concentrate the IND in built-up sections reduced the promotional possibilities on the outskirts. Second, the automobile began to open larger areas for settlement. The realtors thus benefited from the shift to automobiles. Unlike the densely populated corridors spawned by mass transit, the automobile caused a spread-out pattern of growth, which expanded the total amount of land that realtors could sell. This critical shift in land-use development, combined with businessmen's resentment of Mayor Hylan's strident populist rhetoric, soured realtors on rapid transit. Beginning in the 1920s realtors increasingly saw the subways in wholly negative terms, as an unaffordable extravagance that inflated their tax bills and hurt the urban economy.[14] In March 1925, Stewart Browne, president of the United Real Estate Owners' Association, complained that New York's politicians would gladly "throw real estate owners and business interests to the wolves" for a few votes.[15] In March 1942, Henry J. Davenport of the Downtown Brooklyn Association told a meeting of twenty-four business groups that the municipality's transit policy of "soak the property owners" would lead New York "to further trouble, to eventual collapse and thus to chaos."[16]

Paul Windels advocated a ten-cent fare. He calculated in 1943 that a ten-cent fare would generate enough revenue to cover the debt service, reduce the tax rate by twenty-three points, and provide the municipal treasury with $10 million annually.[17] Windels also claimed that the rate increase would lead to the modernization of the subway. Trying to win over straphangers who were weary of wartime overcrowding and of the old, broken-down IRT and BMT trains, Windels pledged that a ten-cent fare would go to buy new aluminum or stainless-steel cars with air conditioning and to transform the dreary stations into well-lit, clean, and cheerful marvels.[18] "Isn't it worth the difference," he asked New Yorkers in a 1943 radio broadcast, "to pay an honest fare instead of a political fare and get decent service instead of the poor service we're getting today?"[19]

Liberals and leftists opposed this business scheme to raise the fare. Perhaps the five-cent fare's ablest defender was Stanley M. Isaacs, a liberal Republican councilman and former Manhattan borough president who was renowned for his strict integrity and his passion for political reform. Rejecting the Committee of Fifteen's view of the subways as a self-supporting business, Isaacs praised rapid transit as a vital municipal service that gave the city "a great unifying force" and that served as the "highway of the masses." Isaacs observed that the city did not charge user fees for the schools, sanitation pickups, fire protection, water supply, or street repairs; and he urged that the same policy be applied to rapid transit through the retention of the nickel fare.[20] Disagreeing with Windels' proposal for the fare to cover capital and operating costs, Councilman Isaacs argued that carrying charges should continue to be supported by the property tax, rather than being shifted to the straphangers. "It should be remembered," he said, "that those who want it changed are the very real estate men who profited most by unloading their property because, as they advertised, it was within reach of the very heart of Manhattan for five cents."[21] Isaacs, who practiced real estate law himself, declared that "the capital charges of those subways which had built up real estate values should be carried by real estate."[22]

Stanley M. Isaacs accused his business adversaries of trying to raise the fare in order to finance the construction of highways. What was involved in this debate, Isaacs claimed, was not making the subways self-sustaining but rather subsidizing a competing form of transportation, the private automobile. In 1945 master builder Robert Moses unveiled an $82 million plan for more than two hundred miles of highways and bridges, including the Throgs Neck and Verrazano-Narrows bridges and the Cross-Bronx, Van Wyck, Brooklyn-Queens, Prospect, and Major Deegan expressways. As Robert A. Caro explained in *The Power Broker: Robert Moses and the Fall of New York* (1975), Moses wanted to fund these projects with the city credit that would be freed once the subway fare covered capital expenses.[23] Isaacs objected, saying that such a shift in capital spending from mass transit to the highways would be unfair to poor and working-class New Yorkers:

> I believe that this is probably the most audacious proposal yet made to saddle those least able to afford it with the cost of civic improvements which in the main serve those in the higher income brackets; to make those who do not own cars but travel in the subway indirectly subsidize the motorist. The whole program is clear. The straphanger is to pay double the present fare so as to carry the full interest upon and amortization of the capital cost of the subways. Why? So that the city will be able to borrow more money to build parkways, expressways, and highways, which are to be furnished free of charge for the capital improvements to the man who can afford his own car, doesn't travel on the subways and doesn't pay even a nickel toward the construction of the speedways furnished him.[24]

When Councilman Isaacs made this statement in 1946, two of every three New Yorkers belonged to households that did not have a car. They would not be able to use the superhighways financed with the released city credit.[25]

The Committee of Fifteen and the Citizens' Transit Committee initially made little headway in their struggle for the ten-cent fare. As long as the subways' fiscal difficulties stemmed from the capital debt, the committees could not overcome the strong popular attachment to the nickel fare. "When the deficit came only from the ability of the system to earn the debt service on the city's transit debt," Windels observed, "the campaign to get a self-sustaining fare progressed slowly."[26] But the New York Transit System's operating surpluses were shrinking. Even though revenues remained high because of the heavy wartime traffic, subway costs were escalating, primarily due to transit workers' wage increases. From 1941 to 1945 wages rose 27 percent, and labor's share of total operating costs went from 54 percent to 63 percent. The New York City Transit System's annual operating surplus went from $27 million in 1941 to $13 million four years later; in 1947 the system sustained its first operating deficit, $18 million.[27]

Although these operating deficits greatly strengthened Windels' case for an increase, New York City's liberals, socialists, communists, and labor union leaders remained firmly committed to the low fare. "It is part of the city itself," communist Councilman Peter V. Cacchione wrote in 1945.[28] But the liberal-left coalition was split by the defection of President Mike Quill of the Transport Workers Union (CIO). Formed by communist organizers in the early 1930s, the Transport Workers Union (TWU) had been a fixture in the city's left-wing politics for over a decade. Mike Quill, the TWU's brilliant and combative Irish-born leader, was widely rumored to be a member of the Communist party. By 1948, however, Quill was moving rightward, prompted by the Cold War and by internal union disputes. To solidify the TWU's position as the bargaining agent for subway workers and to ensure his own mastery of the TWU, Quill broke with longstanding CIO policy and came out in favor of the ten-cent fare in March 1948. He told rank-and-file TWU members that the rate hike would increase their wages.[29] With his left flank protected, Mayor William O'Dwyer announced on April 20, 1948, that the fare would go to ten cents on the subways and elevateds and to seven cents on the streetcars and buses, effective July 1.

Mayor O'Dwyer's decision to raise the fare demonstrated one of the chief consequences of unification: the elimination of the Interborough Rapid Transit Company and the Brooklyn-Manhattan Transit Corporation as political scapegoats. Leaders such as John F. Hylan, Al Smith, and Fiorello H. LaGuardia had curried favor with the voters by lambasting the despised IRT and BMT. Now that the New York City Board of Transportation managed the subways, however, politicians were blamed for transit problems. When William O'Dwyer abandoned the five-cent fare, his mail was running an intimidating fifty-two to one against the increase.[30]

Seizing upon the growing dissatisfaction with political leadership of transit, Paul Windels redoubled his efforts to overhaul subway management.

The Transit Authority

Paul Windels regarded the ten-cent fare as only the first step in curbing transit spending. He thought that the ultimate solution to the subway imbroglio was the creation of a public authority that would end direct municipal operation.

An authority, or special district government, was a public corporation that was autonomous of other branches of government and insulated from popular pressures. The first special district government created in the United States was the Metropolitan Police District, formed by New York State in 1857 to remove the city's police from the clutches of the Tammany machine. A great many public authorities were subsequently established throughout the country, including Boston's Metropolitan Park Authority, New Jersey's Passaic Valley Sewerage Commission, the Chicago Sanitary District, and California's East Bay Utilities District.[31]

Windels' model for reorganizing the city's railways was perhaps the most successful special district government ever established, the Port of New York Authority. Organized in 1921 by New York and New Jersey to rationalize the chaotic metropolitan harbor facilities, the Port Authority presided over the area within twenty-five miles of the Statue of Liberty. This sprawling region had to be administered by a public authority because it encompassed scores of counties, cities, towns, and townships in two states; an ordinary regulatory commission would not have had the flexibility to cross so many jurisdictional lines. Although originally confined to docks and terminals, Port Authority won its financial independence in 1931 by obtaining control of the Holland Tunnel's revenues. It employed its solid bond ratings to become—in the words of Michael N. Danielson and Jameson W. Doig—"the most influential public development institution in the New York region."[32] By the 1940s the Port Authority operated six bridges and tunnels between New York and New Jersey, including the three major crossings of the Hudson River—

the Holland Tunnel, the Lincoln Tunnel, and the George Washington Bridge.[33]

Paul Windels had served as assistant counsel of the Port Authority from 1930 to 1933, and he deeply admired its objectives of technocratic efficiency and professional expertise.[34] Long after resigning from the Port Authority to become corporation counsel in 1934, Windels continued to pay homage to its "great success" in public works construction and regional planning.[35] He believed that the Port Authority exemplified the type of organization that should run the subways. Contrasting the Port Authority's achievements with the Board of Transportation's supposed failures, Windels claimed that the practice of deciding weighty policy issues in an open forum jeopardized public interest instead of serving it. He thought that municipal control would put the subways at the mercy of shortsighted politicians who played to the fickle crowd. He asserted that a public authority staffed by nonpartisan experts would restore a businesslike approach to rapid transit. Windels also wanted a transit authority to stop the Transport Workers Union from acquiring more power, partly because higher wages accounted for much of the general cost increases and partly because he disliked the union's leftist ideology. He later maintained that "control over operations has been lost since 1946 to Michael Quill of the Transport Workers Union, C.I.O."[36]

For Windels and his conservative supporters, "politics" had a clear and exclusive meaning: the maneuvering of professional officeholders in party clubhouses and electoral races. Theirs was a limited point of view that disregarded legislative lobbying, "good government" campaigns, fundraising, and a host of other methods by which influential groups—including themselves—attempted to make their voices heard. Instead, Windels and the businessmen who led this revolt against politics focused on eliminating the citizenry's principal mechanism for changing policy decisions—the vote.

Paul Windels' proposal for a public authority was a response to transit's basic political difficulties, especially the simultaneous popular demands for a cheap fare and high passenger standards. He realized that the subways' problems were primarily political in origin rather than economic or structural; periodic financial crises such as postwar inflation revealed strains in the political system and created an impetus for change. So Windels sought to end the convulsions over the nickel fare, the IND, and unification that had shaken New York City ever since the days of Mayor John F. Hylan. He wanted to strip the subways away from the mayor and the city council. Aware that the political system had been effective in communicating New Yorkers' desire for a low-fare subway, Windels resolved to change the system. By taking transit "out of politics," he hoped to guarantee that there would be no more wide-open disputes over subway issues, no more resistance to the imposition of fiscal discipline. By establishing a transit authority, Windels intended to change the subways' political culture and ensure the triumph of his ideology of business management.[37]

Although business groups such as the Citizens Budget Commission and the Queens Chamber of Commerce endorsed Windels' campaign for a transit authority, liberals and leftists recoiled from it.[38] But the left-liberal effort to justify public subsidies was compromised by the LaGuardia administration's definition two decades earlier of mass transit as a business enterprise. LaGuardia had been convinced that unification would provide great operating economies, so he and his aides assumed that the social benefits of a nickel fare were compatible with good business management. The ambiguity of LaGuardia's stance became apparent only in the 1940s and 1950s when annual operating deficits made the need for aid painfully clear. In the politically more conservative and automobile-oriented postwar climate, however, subway defenders had little hope of prevailing.[39]

The subways also suffered from a political anomaly. New York was perhaps the only world-class city that was not a national capital, and its subway was the only major one not located in a capital city. Unlike the London, Paris, Berlin, Moscow, and Tokyo metros, New York's subway had not been conceived as a national showcase and did not receive extra funding for reasons of prestige. The contrast between New York City and London was especially marked. Like the New York subway, the London underground consisted of two private subway companies—the Underground Group Combine and the Metropolitan Railway—that underwent a financial crisis in the early 1930s. In 1933, Parliament approved a measure, called the London Passenger Transport bill, that unified the capital's undergrounds, buses, and trams under the auspices of a public corporation known as London Transport. Although London Transport was supposed to be self-sustaining, the underground fared better than the New York subway, at least until Great Britain's economic collapse finally took its toll during the 1980s. There was substantial public investment in the physical unification of the underground's separate networks and in the construction of the new Victoria line, which opened in 1969. This high level of spending was due partly to London's status as a national capital. Without a corresponding role as a national symbol that might have overcome the federal government's reluctance to invest in local public works, the New York subway was forced to rely on the hard-pressed state and municipal governments through the 1940s and 1950s.[40]

Paul Windels' business definition of rapid transit easily triumphed over the flawed liberal visions of the subways. Support for Windels' transit authority proposal grew in the early 1950s when the subways again encountered serious financial difficulty. The adoption of the ten-cent fare proved to be just a stopgap; although the New York City Transit System made a surplus in fiscal year 1948, patronage continued to fall and labor costs kept climbing. The transit system incurred operating deficits of $1.2 million in 1950 and $3.1 million in 1951. It lost an alarming $24.8 million in 1952, and some experts were predicting that it would lose $50 million the following year. These tremendous shortfalls appeared to confirm Paul Windels' argument that a transit authority was needed to free the subways from political interference and "to recover disciplinary control over the organization and stop waste and inefficiency."[41] By 1952 powerful business groups such as the Commercial and Industry Association and the Board of Trade and Transportation were pushing for a transit authority.[42]

On March 10, 1953, Republican Governor Thomas E. Dewey called on the legislature to create such an authority. Dewey argued that an authority would remove transit charges from the city's capital budget and ensure that the subways no longer competed for scarce resources with schools, hospitals, highways, and other services. He held that an authority would foster "genuinely independent" and "business-like" administration which would improve passenger service and produce operating economies.[43] Dewey's proposal was quickly introduced into the senate and the assembly.[44]

Liberals and leftists abhorred the governor's plan. The United Department Store Workers (CIO) warned that "by this means the power to control and limit transit fares passes from the hands of the people of the city to a political body not responsible to the electorate."[45] F. W. Grumman, secretary of Local 100 of the American Communication Association, feared that the authority "would mean the shifting of the tax burden onto the shoulders of the working population."[46]

But the Republicans controlled both the senate and the assembly, and the proposal sailed through the legislature.[47] Governor Dewey signed the bill on March 20, 1953. He promised that the new authority would make the subway "the greatest transit system in the world" and ensure "efficient management and the elimination of politics from its operation."[48]

The New York City Transit Authority (NYCTA), which leased the rapid transit railway from the City of New York for a period of ten years, began to operate on June 15, 1953. The NYCTA was a semi-independent public corporation run by a five-member board of directors; the governor and the mayor would each select two members, and their appointees would name the fifth member. According to Paul Windels, these directors would exercise virtually total control over the subways, including the right to raise the fare.[49] As an independent corporation designed to be free of city hall and removed from electoral politics, the NYCTA was supposed to have great power over the subways.[50]

One critic, however, had already questioned Windels' claim that a transit authority would be a panacea for the subways. In a scathing 1949 memo, Robert Moses insisted that the subways would not generate enough revenue for a transit authority to be genuinely independent. "We believe that the 'authority' method of financing is valid in several fields of public construction and management," Moses asserted, "but that gold bricks wrapped up as bullion will kill this device." He concluded that "such an authority is not practical" for the subways and that a transit authority would probably not repeat the success of the Port Authority or his own Triborough Bridge Authority.[51]

Moses was right. Rather than curing rapid transit's existing flaws, Paul Windels' transit authority proposal aggravated the situation. The NYCTA, like so many other special district governments, was a highly bureaucratic organization that became largely unanswerable to the public.

Yet more was involved than just the type of government structure that controlled New York's subways, elevateds, trolleys, and buses; the NYCTA was by no means the villain responsible for the subway's decline. Instead, it was a response to more general weaknesses in the political culture of the subways. At root, the NYCTA was established because city leaders did not want to deal with the subways' problems. This desire to turn a blind eye to the subways was hardly new in 1953. From Mayor John F. Hylan's demogogic assaults on the BMT and IRT, through Mayor Fiorello H. LaGuardia's shortsighted conception of unification, to Mayor William O'Dwyer's pained acceptance of the ten-cent fare, New York City's leaders tried to avoid dealing with subway issues. The combination of strong popular demands for a low-fare subway and the ideology of business management made subway politics extremely difficult. New Yorkers and their leaders failed to formulate a new understanding of the subways as a municipal enterprise or to overcome the deep-seated opposition to public investment in rapid transit. The problem clearly extended beyond New York City's borders. At a time when the federal government was subsidizing the growth of suburbs through the establishment of low-interest mortgage insurance and the construction of highways, rapid transit continued to be conceived in local rather than national terms and to be defined as a business enterprise. The NYCTA thus represented the culmination of three decades of political failures.[52]

The NYCTA ushered in a new, more bureaucratic political system that enshrined the ideology of business management and that insulated subway administration from the general public. How spectacularly ill-suited this new political system was to meeting rapid transit's needs became clear in the 1960s and 1970s when inflation, rising operating costs, and decreasing ridership reduced transit's real income once again and saddled the NYCTA with mammoth operating deficits. The political culture rendered the authority incapable of responding to this latest fiscal crisis. By insulating the subways from standard governmental channels and lessening the accountability of the governor, the mayor, and other top officials, the creation of the NYCTA further weakened the public commitment to rapid transit. By making economy the highest priority, the NYCTA deprived officials of the taxing powers needed to put this deficit-plagued system on a solid footing and avert physical collapse. By excluding the general public and many private groups from decisions, the NYCTA discouraged people from thinking of the subways as a civic enterprise and trying to change public policy.[53]

The creation of the NYCTA set the stage for the subways' deterioration in the 1970s and afterward. Ridership continued to drop as more New Yorkers took to the highways, and the fare had to be raised ten times between 1953 and 1990. Nine of every ten subway trains had run on schedule during the 1930s and 1940s, but by 1983 on-time performance fell to 70 percent. The average distance a subway car traveled between breakdowns dropped from 34,294 miles in 1964 to 9,000 miles in 1984. Rising crime and graffiti made the system a national symbol of urban violence and disorder, especially in movies such as *The Incident* (1967) and *The Taking of Pelham One Two Three* (1974). By any measure the subways had hit bottom.[54]

EPILOGUE

THE KITCHEN DEBATE

New York City and Moscow had little in common in the 1950s. One was the home of Wall Street and the engine of the most powerful economy in the world; the other was the capital of the Soviet Union and the wellspring of communism. For six weeks during the summer of 1959, however, a piece of the New York suburbs arose in Moscow's Sokolniki Park.

At the time, the United States and the Soviet Union were experiencing a thaw in the Cold War. To ease international tensions and encourage mutual understanding, the two superpowers had decided to mount a cultural fair in Moscow, called the American National Exhibition, to show the Russian people how ordinary Americans lived, worked, and played. Organized by the U.S. Departments of State and Commerce and dominated by large corporations, the exhibition communicated an idealized image of American society that stressed the benefits of consumer capitalism and suburbia. It featured late-models Fords and Chryslers, RCA color televisions, Marantz and Fisher hi-fis, Polaroid cameras, Whirlpool's "Kitchen of the Future," and a Pepsi-Cola stand. The American National Exhibition also included a unique living display that dramatized the advantages of postwar America. The fair's organizers had selected the R. Ted Davises of Millburn, New Jersey, as the "Typical American

Suburban Family." The Davis family consisted of Mr. Davis, an advertising salesman for an architectural magazine in New York who served as an usher at Christ Episcopal Church and a Boy Scout leader in Millburn; Mrs. Davis, a homemaker, Sunday school teacher, and PTA volunteer; and their three children: Jeff, fifteen years old; Jane, twelve; and Chuck, ten. Flown to Moscow for the exhibition, the Davises symbolized the sponsors' perception of the best of American society.[1]

On July 24, 1959, Soviet Premier Nikita S. Khrushchev and U.S. Vice-President Richard M. Nixon toured the American National Exhibition and had one of the most celebrated encounters of the Cold War, the kitchen debate. The debate took place in a "typical" American ranch house built by All-State Properties of suburban Long Island; it was similar to the tens of thousands of homes that were being constructed throughout the United States during the 1950s. By using a government-subsidized mortgage from the Veterans Administration or the Federal Housing Authority, a buyer could purchase this six-room house for $13,000, exclusive of land, and pay it off over twenty-five or thirty years. This low price was within reach of many semi-skilled workers. Indeed, All-State boasted that this inexpensive ranch house was a secret weapon in the Cold War because it gave ordinary Americans a high standard of living and inoculated them against the contagion of radical ideas.

Premier Khrushchev and Vice-President Nixon saw this suburban house in competitive terms, too. As they walked through it, Khrushchev and Nixon began arguing heatedly about the relative merits of the United States and the Soviet Union. They ended up in the kitchen, and most of their impromptu debate took place there. Khrushchev ridiculed the kitchen appliances—such as an automatic lemon squeezer that took more time to use than a hand one—as worthless gadgets that were completely impractical. But Nixon was on his own grounds in this kitchen. Fully equipped with a General Electric oven and range, a refrigerator, and an automatic washing machine that was built into the wall, the kitchen embodied the American dream of individual social mobility, private family life, and material consumption. Looking intently at the washing machine crowned with a box of Dash detergent, Nixon proudly told Khrushchev that almost any American steelworker could purchase such a house. To Vice-President Nixon, the superiority of the West was clear.

Most Americans who read about the debate in their newspapers or in *Time* magazine or who saw the famous photograph of Nixon jabbing his finger in Khrushchev's chest had no doubt about who had "won" the debate.[2] As historian Michael R. Beschloss observed, the kitchen debate "soon entered American folklore" as a ringing vindication of capitalism and democracy.[3]

But there was a postscript to the kitchen debate that escaped public notice. Soviet officials evidently agreed that Khrushchev had "lost" the argument, and they wanted another opportunity to develop his critique of American society. The next day First Deputy Premier Frol R. Kozlov spoke to a group of American reporters and leveled an angry attack on the United States, blasting it as a spiritually bankrupt culture that excelled at producing consumer gimmicks and restricted economic success to the privileged few. The deputy premier had recently visited the United States and had acquired a working knowledge of the country. Kozlov thus gave one example that illustrated America's social ills: the New York City subway.

From Kozlov came the Kremlin's verdict on the IRT, BMT, and IND: "lousy." He complained that "the subways are dirty and the air is bad—very bad." Asked how New York's underground could be improved, Kozlov replied that it was beyond repair. "You would just have to reconstruct it, I think."[4] Kozlov's choice of the subway as a means of criticizing American society was inspired; New York's underground railway was an apt symbol of the older, more urban and industrial society that the organizers of the American National Exhibition had neglected in their eagerness to celebrate suburbia and consumerism.

The contrast between the New York subway and the Moscow metro was especially striking. Many of the 3.7 million people who visited the American National Exhibition took the metro to the fairgrounds in Sokolniki Park, which was located fifteen minutes by subway from downtown Moscow. Begun in the 1930s under Joseph Stalin's personal leadership and planned as a showcase for his regime, the Moscow metro was the product of a totalitarian government that could concentrate massive social resources on a project which had top national priority. The metro was magnificent; its stations featured chandeliers, stained-glass panels, bronze and ceramic statues, marble walls, and mosaic murals that depicted the achievements of the Soviet Union. Radiating from the center to the outskirts, the metro lines provided inexpensive transportation for a poor population that could not afford private automobiles.[5]

Compared to this Soviet wonder, the New York subway was in bad condition in 1959. The trains were shabby, the stations were dirty, the fare was high (fifteen cents). Even the *New York Times* had to concede that Deputy Premier Kozlov was "a man who knows his subways."[6] New York's subway typified the decline of public street life and the deterioration of the urban infrastructure that had occurred as more and more middle-class residents moved to the suburbs. Even though the United States was much wealthier than the Soviet Union, it lacked the political will or the means required to apply its resources to rapid transit.

The irony, of course, was that the origins of New York's subway were as spectacular as those of the Moscow metro. Conceived by mercantile leaders such as Abram S. Hewitt, Alexander E. Orr, and August Belmont, who had a far-reaching vision of New York as a commercial metropolis that would dominate world trade, the subway was a political masterpiece that combined business and government in an innovative partnership. It was also a gigantic construction and engineering project that overcame severe physical obstacles such as Manhattan's difficult geology and the East River. The subway remade New York City's landscape by stimulating real estate development in upper Manhattan and in the outer boroughs. It changed the way New Yorkers experienced their city by moving passengers below the ground, shortening distances, and quickening the pace of urban life.

But this initial success was not maintained, and the subway began to lose its high standing as early as the 1920s. The rise of inflation during World War I strained the existing arrangements between the City of New York and the two private subway companies. When Mayor John F. Hylan made an issue of the nickel fare in his 1921 election campaign, such important transit matters as the rate of fare and expansion plans became politicized. The nickel fare began a process of disinvestment in the subways that crippled passenger service; in the end, political failures led to the physical deterioration of the subways.

The city failed to create new political arrangements that could have solved the chronic problem of subway financing. The key obstacles were that rapid transit continued to be defined in local rather than national terms, that the combination of strong popular demands for a low-fare subway and for high-quality passenger service created impossible political pressures, and that the ideology of business management precluded government investment in rapid transit. By continuing to define the subways as a self-sustaining enterprise, New Yorkers were unable to formulate a new understanding of underground rapid transit that could justify direct government subsidies. The unification of rapid transit under public control in 1940 did not correct the subway's financial problems, which worsened after World War II due to increased competition from automobiles. The establishment of the New York City Transit Authority in 1953 removed the subways from electoral politics and permitted minimal standards to prevail. This, in turn, set the stage for the severe physical decline of the 1960s and 1970s.

In the early 1980s the NYCTA began to reconstruct the system. It replaced or overhauled every subway car, rebuilt every mile of main-line track, and rehabilitated fifty-six stations. These improvements are a sign of hope in an otherwise bleak picture; however, this major reconstruction has been endangered by the weak economy of the 1990s.[7]

It is hardly surprising that First Deputy Premier Frol R. Kozlov knew nothing about the development of the New York subway. As a Soviet leader who had spent only a brief time in the United States, Kozlov could hardly be expected to be an authority on New York City. What is significant, however, is that New Yorkers failed to challenge First Deputy Premier Kozlov's biting condemnation of their subways. Having little memory of how splendid the subways had once been and no faith that they could be revived, they most probably agreed with Kozlov.

With its brilliant record as a thriving commercial and financial center, New York has always been a city that buried its history and looked confidently to the future. The original IRT subway had been forged in that heady, optimistic spirit. Now, its glory days all but forgotten, the subway belongs to New York's past rather than its future.

ACKNOWLEDGMENTS

I am delighted to acknowledge my enormous debt to Kenneth T. Jackson, the Jacques Barzun Professor of History and Social Sciences at Columbia University. Kenneth T. Jackson directed the doctoral dissertation upon which this book is based, but that plain statement of fact does not do justice to the full scope of his contributions. Ken Jackson is a remarkably good-hearted and generous man, and he has enriched me with his many personal kindnesses, his intellectual guidance, and his wise professional counsel. I feel lucky to know him.

Walter P. Metzger, also of Columbia University, encouraged me during my years in graduate school and carefully followed my dissertation through several phases. At my doctoral defense, Robert A. McCaughey of Barnard College provided tough-minded criticism that forced me to rethink my conceptualization of this subject. I am grateful for his help. Peter Derrick of the Metropolitan Transportation Authority read my entire dissertation and most of these chapters. I relied heavily on Peter's superb knowledge of New York's transit system, and I enjoyed our friendly arguments about subway politics.

Carol Mann, my agent, saw promise in this study at an early stage and made the connection with Simon & Schuster that led to its publication; without her support and commitment, this book would exist only in my imagination. Bob Bender, my wonderful and gifted editor, shared my fascination with New York City and applied his uncanny sense of structure to the text.

I would like to thank Paul Barrett, Lawson H. Bowling III, Evan W. Cornog, Jameson W. Doig, Robert Fishman, Joshua B. Freeman, Timothy J. Gilfoyle, Sigurd Grava, David C. Hammack, Elizabeth Hovey, Ann Durkin Keating, Thomas Kessner, Jeffrey A. Kroessler, Richard K. Lieberman, Chris McNickle, Deborah Dash Moore, Clement A. Price, Daniel J. Singal, Joel A. Tarr, William S. Vickrey, Marc A. Weiss, and Olivier Zunz for reading and commenting on various parts of the book. I profited from Kerry Michaels' ideas and suggestions, her unbounded enthusiasm for my project, and, above all, her close friendship. Vincent F. Seyfried, who knows more about the history of Queens than anyone, graciously opened his private research files on Jackson Heights and the Steinway tunnel. Mary Elizabeth Brown, Annie Chamberlin, and Mary Curtin helped me with the research.

I also want to express my deep appreciation to the university seminars at Columbia University for assistance in the preparation of the manuscript for publication. The ideas presented have benefited from discussions in the university seminar on the city.

I would like to thank my friend Abigail Sturges, who designed the book, and Rose Ann Ferrick, who copyedited it. I am also grateful to Gypsy da Silva, for guiding the book through production, and Johanna Li, for moral support.

I owe special thanks to Kenneth R. Cobb, the director of the Municipal Archives of the New York City Department of Records and Information Services. Ken was the first friend I made in New York City. He shepherded me through the Municipal Archives' rich collections and answered my seemingly endless research questions with patience and good cheer; perhaps more important for my sanity, Ken was always willing to grab a pizza or see a movie. I would also like to salute the staffs of the New York Public Library, Columbia University libraries, New-York Historical Society, Seely G. Mudd Manuscript Library at Princeton University, Museum of the City of New York, Queens Borough Public Library, University Libraries of

Pennsylvania State University, Franklin D. Roosevelt Library, Rush Rhees Library of the University of Rochester, Electric Railroaders Association, Minnesota Historical Society, New York Transit Museum, and my former colleagues at LaGuardia Community College/CUNY.

NOTES

In these notes, manuscript collections and government agencies that are cited frequently are abbreviated as follows:

BFP: Belmont Family Papers, Rare Book and Manuscript Library, Butler Library, Columbia University, New York, NY. The Belmont Family Papers contain two kinds of letterbooks, private letterbooks and confidential letterbooks. Private letterbooks are abbreviated PL and confidential letterbooks are abbreviated CL.

FHL: Fiorello H. LaGuardia Papers, Municipal Archives, New York City Department of Records and Information Services, New York, NY.

JHHP: John H. Hendrickson Papers, Long Island Division, Queens Borough Public Library, Jamaica, NY.

NYCBT: New York City Board of Transportation.

NYCRTC: New York City Board of Rapid Transit Railroad Commissioners (created by the act of 1894).

NYCTA: New York City Transit Authority.

NYSPSC: New York State Public Service Commission for the First District.

NYSTC: New York State Transit Commission.

S&SP: Steinway & Sons Papers, LaGuardia and Wagner Archives, LaGuardia Community College/CUNY, Long Island City, NY. The Diary of William Steinway is abbreviated as SD.

SMI: Stanley M. Isaacs Papers, Division of Rare Books and Manuscripts, New York Public Library, New York, NY.

Introduction

1. New York City Transit Authority, *New York City Transit Authority's Facts and Figures* (New York: n.p., 1991): unpaginated. Boris S. Pushkarev with Jeffrey M. Zupan and Robert S. Cumella, *Urban Rail in America: An Exploration of Criteria for Fixed-Guideway Transit* (Bloomington, IN: University of Indiana Press, 1982): table A-2. *Jane's Urban Transport Systems* (London: Jane's Publishing Company, 1984): 366. U.S. Department of Justice, Federal Bureau of Investigation, *Crime in the United States, 1987* (Washington, D.C.: U.S. Government Printing Office, 1987): 235–96.

2. Metropolitan Transportation Authority, *A Methodology Report on the Comprehensive Travel Telephone Survey* (New York, 1990): 4.

3. Roger Starr, "New York's Singular Subway," *The City Journal* 2 (Winter 1992): 2–3. New York City Transit Authority, *New York City Transit Authority's Facts and Figures:* unpaginated. *New York Times,* April 30, 1992. Interview with Mark Hess, NASA spokesman, May 1, 1992. Though not nearly as large as the New York subway, rapid transit systems are, of course, now being constructed with federal aid in cities such as Los Angeles, Atlanta, and Miami. For the rise of the federal role with the Urban Mass Transit Act of 1966, see Alan Altshuler with James P. Womack and John R. Pucher, *The Urban Transportation System: Politics and Policy Innovation* (Cambridge, MA: MIT Press, 1981): 32–37, 301–4.

4. The best general account of New York City's rapid transit system remains James Blaine Walker, *Fifty Years of Rapid Transit, 1864–1917* (New York: Law Printing Company, 1918). For more specialized studies, see Charles W. Cheape, *Moving the Masses: Urban Public Transit in New York, Boston, and Philadelphia* (Cambridge, MA: Harvard University Press, 1980); David C. Hammack, *Power and Society: Greater New York at the Turn of the Century* (New York: Russell Sage Foundation, 1982): 230–58; Joshua B. Freeman, *In Transit: The Transport Workers Union in New York City, 1933–1966* (New York: Oxford University Press, 1989); Peter Derrick, "The Dual System of Rapid Transit: The Role of Politics and Subway Planning in the Second Stage of Subway Construction" (Ph.D. dissertation, New York University, 1979); Cynthia M. Latta, "The Return on the Investment of the Interborough Rapid Transit Subway" (Ph.D. dissertation, Columbia University, 1975); Arthur J. Waterman, "The Integration of Rapid Transit Facilities of the City of New York" (Ph.D. dissertation, New York University, 1940); and Joel Fischer, "Urban Transportation: Home Rule and the Independent Subway System in New York City, 1917–1925" (Ph.D. dissertation, St. John's University, 1975). The best popular history of the subway is Brian J. Cudahy, *Under the Sidewalks of New York: The Story of the Greatest Subway System in the World,* rev. ed. (Lexington, VT: Stephen Greene Press, 1979). Another good popular account is Jim Dwyer, *Subway Lives: A Day in the Life of the New York City Subway* (New York: Crown, 1991).

5. The Transit Authority inaugurated a third, more bureaucratic period that began in 1953. Although this period is beyond the scope of this study, it has

possibilities for further research. Its major components include the rivalry between the Transit Authority and the Transport Workers Union, the advent of a new relationship with the federal government, the legitimation of limited government subsidies, widespread racial animosities, increased crime, and a growing perception of disorder. In addition to investigating the decline of the subways, however, researchers should also consider the reconstruction of the system that began in the early 1980s as subway cars were replaced or reconditioned, as mainline track was rebuilt, and as stations were rehabilitated.

6. See Paul Barrett, *The Automobile and Urban Transit: The Formation of Public Policy in Chicago, 1900–1930* (Philadelphia: Temple University Press, 1983); Glen E. Holt, "The Changing Perception of Urban Pathology: An Essay on the Development of Mass Transit in the United States," in *Cities in American History,* ed. Kenneth T. Jackson and Stanley K. Schultz (New York: Alfred A. Knopf, 1972): 324–43; Glenn Yago, "The Decline of Public Transit in the United States and Germany, 1900–1970" (Ph.D. dissertation, University of Wisconsin, 1980); T. C. Barker and Michael Robbins, *A History of London Transport,* 2 vols. (London: George Allen and Unwin, 1975 and 1976); Donald Dewess, "The Decline of the American Street Railways," *Traffic Quarterly* 24 (October 1970): 563–81; Stanley Mallach, "Origins of the Decline of Urban Mass Transportation in the United States, 1890–1930," *Urbanism Past and Present* 8 (Summer 1979): 1–17; Mark S. Foster, *From Streetcar to Superhighway: American City Planners and Urban Transportation, 1900–1940* (Philadelphia: Temple University Press, 1981); David J. St. Clair, "The Motorization and Decline of Urban Public Transit, 1935–1960," *Journal of Economic History* 41 (September 1981): 589–600.

Prologue: Abram S. Hewitt

1. Chamber of Commerce of the State of New York, *Forty-fourth Annual Report, 1901–1902* (New York: Press of the Chamber of Commerce, 1902): 26–32. *New York Times,* February 1, 1901.

2. Frederic Cople Jaher, *The Urban Establishment: Upper Strata in Boston, New York, Charleston, Chicago, and Los Angeles* (Urbana, IL: University of Illinois Press, 1982): 159–60, 178–96. Allan Nevins, *Abram S. Hewitt with Some Account of Peter Cooper* (New York: Harper & Brothers, 1935): 3–10. *New York Times,* January 19, 1903. *Dictionary of American Biography,* v. 8, ed. by Dumas Malone (New York: Charles Scribner's Sons, 1932): 604–6. *National Cyclopaedia of American Biography,* v. 3 (New York: James T. White and Company, 1893): 294–95.

3. Nevins, *Abram S. Hewitt:* 461–69. David Scobey, "Boycotting the Politics Factory: Labor Radicalism and the New York City Mayoral Election of 1884 (sic)," *Radical History Review* 28–30 (Fall 1984): 280–325.

4. William R. Grace served two nonconsecutive terms, 1881–82 and 1885–86.

5. Nevins, *Abram S. Hewitt:* 471–91. Abram S. Hewitt to George C.

Ohren, January 25, 1888, Mayoral Book 1, Hewitt Mayoral Papers, New-York Historical Society, New York City, NY.

6. Abram S. Hewitt, *Message to the Board of Aldermen* (New York: n.p., 1888): 5, 15, 19, 27, 35.

7. NYSPSC, *Report, 1919,* v. 2 (Albany: n.p., 1920): 12–17. U.S. Department of the Interior, Census Office, *Report on Transportation Business in the United States at the Eleventh Census; 1890,* Part 1, *Transportation by Land* (Washington, D.C.: Government Printing Office, 1895): 688, 709–10. The Third Avenue Railway Company also operated a short cable line; the total length of cable railway in New York was 6.89 miles.

8. Hewitt, *Message to the Board of Aldermen:* 31–32. Hewitt wanted the New York Central Railroad to operate his route. The Central, which already controlled the trunk railroads into Manhattan, would have thoroughly dominated regional transportation if this plan had been carried out. Hewitt did not specify whether an elevated or a subterranean railway should be built, although he did make it clear that an elevated railway, if built, would have to be superior to the existing els.

9. Ibid.: 37.

10. Patrick Hughes, *American Weather Stories* (Washington, D.C.: U.S. Government Printing Office, 1976): 60–75.

11. *New York Times,* February 12, 19, and 22, 1888.

12. Chamber of Commerce, *Forty-fourth Annual Report:* 27, 32.

13. Ibid.: 27, 32.

Chapter 1. The Great City

1. Cadwallader D. Colden, *Memoir, Prepared at the Request of the Committee of the Common Council of the City of New York, and Presented to the Mayor of the City, at the Celebration of the Completion of the New York Canals* (New York: n.p., 1825): 64, 189–99. I. N. Phelps Stokes, *The Iconography of Manhattan Island,* v. 5 (New York: Robert H. Dodd, 1926): 1651–52.

2. Robert Greenhalgh Albion, *The Rise of New York Port, 1815–1860* (New York: Charles Scribner's Sons, 1939; reprint ed., Boston: Northeastern University Press, 1984): 235–58. *National Cyclopaedia of American Biography,* v. 1 (New York: James T. White and Company, 1898): 500–1. *National Cyclopaedia of American Biography,* v. 6 (New York: James T. White and Company, 1929): 208–9.

3. Frederic Cople Jaher, *The Urban Establishment: Upper Strata in Boston, New York, Charleston, Chicago, and Los Angeles* (Urbana, IL: University of Illinois Press, 1982): 173–84, 222–37, 245–50.

4. *New York As It Is, in 1833, and Citizens Advertising Directory* (New York: J. Disturnell, 1833): 8.

5. Albion, *The Rise of New York Port:* 39–47, 95–121, 178–93, 198–205.

6. Ellis L. Armstrong, ed., *History of Public Works in the United States, 1776–1976* (Chicago: American Public Works Association, 1976): 26. Dorothie Bobbe, *DeWitt Clinton,* 2nd ed. (Port Washington, NY: Ira J. Friedman, Inc.,

1962): 145–64, 273–76. For the impact of the Erie Canal on upstate New York, see Whitney R. Cross, *The Burned-over District: The Social and Intellectual History of Enthusiastic Religion in Western New York, 1800–1850* (New York: Harper & Row, 1965): 56–59, 154–55, 233, and Paul E. Johnson, *A Shopkeeper's Millenium: Society and Revivals in Rochester, New York, 1815–1837* (New York: Hill & Wang, 1978): 15–16, 105–6.

7. Albion, *The Rise of New York Port:* 386, 418.

8. Stokes, *Iconography of Manhattan Island,* v. 5: 1651.

9. Ira Rosenwaike, *Population History of New York City* (Syracuse, NY: Syracuse University Press, 1972): 16, 36, 58–59, 63. Population figures for the greater city have been used for the sake of comparison.

10. Norval White and Elliot Willensky, *AIA Guide to New York City,* rev. ed. (New York: Collier Books, 1978): 576–83.

11. Rosenwaike, *Population History of New York City:* 36. Charles Lockwood, *Manhattan Moves Uptown* (Boston: Houghton Mifflin Company, 1976): 3–6. Donald A. McKay, *The Building of Manhattan* (New York: Harper & Row, 1987): 54–56. James Hardie, *The Description of the City of New York* (New York: Samuel Marks, 1827): 146. Robert Sutcliffe, *Travels in Some Parts of North America in the Years 1804, 1805 & 1806* (London: C. Peacock, 1911): 25. *New York Herald-Tribune,* March 12, 1950. *New York Times,* September 9, 1988.

12. John H. White, Jr., *Horse Cars, Cable Cars and Omnibuses* (New York: Dover Publications, 1974): vii–xviii.

13. Ibid: vi–xvii. George Rogers Taylor, "The Beginnings of Mass Transit in Urban America: Part 1," *Smithsonian Journal of History* 1 (Summer 1966): 40–47. Glen E. Holt, "The Changing Perception of Urban Pathology: An Essay on the Development of Mass Transit in the United States," in *Cities in American History,* ed. Kenneth T. Jackson and Stanley K. Schultz (New York: Alfred A. Knopf, 1972): 324–27. Stokes, *Iconography of Manhattan Island,* v. 5: 530–31.

14. George Augustus Sala, *My Diary in America in the Midst of War,* v. 1 (London: Tinsley Brothers, 1865): 87.

15. Allan Nevins, ed., *The Diary of Philip Hone, 1828–1851* (New York: Dodd, Mead and Company, 1927): 730.

16. Lockwood, *Manhattan Moves Uptown:* 51.

17. Samuel B. Halliday, *The Little Street Sweeper; or, Life Among the Poor* (New York: Phinney, Blakeman, & Mason, 1861): 207–9. Matthew Hale Smith, *Sunshine and Shadow in New York* (Hartford, CT: J. B. Burr and Company, 1869): 202–13. Citizens' Association of New York, *Report of the Council of Hygiene and Public Health of the Citizens Association of New York upon the Sanitary Condition of the City,* 2nd ed. (New York: D. Appleton and Company, 1866): lxxi–lxxv. Robert Ernst, *Immigrant Life in New York City, 1825–1863* (New York: King's Crown Press, 1949; reprint ed., New York: Octagon Books, 1979): 17–19.

18. NYSPSC, *Report, 1919,* v. 2 (Albany: n.p., 1920): 13. *Miller's New York As It Is; or Stranger's Guide-Book to the Cities of New York, Brooklyn, and Adjacent Places* (New York: James B. Miller, 1866): 99–101.

19. Walt Whitman, *New York Dissected,* ed. Emory Holloway and Ralph

Adimari (New York: Rufus Rockwell Wilson, 1936): 119–21. *Miller's New York As It Is:* 24.

20. Asa Greene, *A Glance at New York* (New York: A. Greene, 1837): 5.

21. *New York Herald,* October 2, 1864.

22. Ibid.

23. Ibid.

24. Ibid. "Rapid Transit in New York," *Appleton's Journal* n.s. 4 (May 1878): 393.

25. *New York Herald,* October 2, 1864. It should be noted that what the *Herald* wanted was a cheap cab system, not rapid transit. This policy preference denoted both the newness of rapid transit and the paper's primary concern with middle-class riders.

26. A. P. Robinson, *Report upon the Contemplated Metropolitan Railroad of the City of New York* (New York: n.p., 1865): 1–6, 10–15. *Articles of Association of the Metropolitan Railway Company* (New York: n.p., 1864): 1–7. *The Underground Railway Project: Who Originated It, and Who Have Been Its Chief Promoters?* (New York: n.p., 1867): 1–13. For Beach's concern about political corruption, see Alexander B. Callow, *The Tweed Ring* (New York: Oxford University Press, 1965): 185–88.

27. *Scientific American,* n.s. 74 (January 11, 1896): 166. *The National Cyclopaedia of American Biography,* v. 8 (New York: James T. White and Company, 1924): 122–23.

28. T. C. Barker and Michael C. Robbins, *A History of London Transport,* v. 1, *The Nineteenth Century* (London: George Allen & Unwin, 1975): 99–103, 113–22.

29. Alfred E. Beach, *The Pneumatic Dispatch* (New York: American News Company, 1868): 7–8, 16, 20, 22–23.

30. Ibid.: 34–38. Although Beach introduced pneumatic power to the United States, he did not open the first pneumatic subway line. In September 1864, T. W. Rammel had unveiled a passenger tube at the Crystal Palace in London, on a quarter-mile route from the Sydenham entrance to Penge-gate.

31. *Proposed Pneumatic Dispatch and Express Company* (New York: Clayton & Medole, 1865): 2–10.

32. Robert Cruikshank, *An Exposition of Facts & Circumstances Relating to the Tunnel under the Thames* (London: n.p., 1833): 2–4. Beach Pneumatic Transit Company, *Illustrated Description of the Broadway Underground Railway* (New York: n.p., 1872): 1–8. See also Benson Bobrick, *Labyrinths of Iron: Subways in History, Myth, Art, Technology, & War* (New York: William Morrow, 1981): 185–88.

33. Beach Pneumatic Transit Company, *Illustrated Description:* 1–12.

34. *New York Herald,* February 27, 1870.

35. "Rapid Transit in New York," *Appleton's Journal:* 395.

36. *To the Friends of Rapid City Transit* (New York: n.p., 1871): 2–24. Henry B. Sedgwick and Nelson J. Waterbury, *History of Underground Charters and Their Present Legal Status* (New York: n.p., 1880): 2–4.

37. Beach, *Pneumatic Dispatch:* 7–8, 16, 20, 22–23. The best-known parcel dispatch system was perhaps the Paris pneumatique. Begun by the French postal ministry in 1867, it eventually totaled 240 miles of tubes. At its high point

between the world wars, it could deliver letters to any part of the city in only two hours. The Paris pneumatique closed in 1984 due to financial losses. *New York Times,* March 31, 1984.

38. One drawback to a shallow-cut trench would have been the disruption of traffic and shopping on Broadway. This would have been a major obstacle, especially in light of the fact that department store owner A. T. Stewart marshaled political opposition to deep tunnels for fear that their construction would undermine structural foundations. Yet the route of the subway could have been changed to a less important peripheral artery such as Greenwich Street or the Bowery; New York City probably could have tolerated the disruption of one of its north-south arteries in return for underground mass transportation. *The Broadway Underground Railway: The Only True Solution of the Rapid Transit Problem* (New York: n.p., 1876): 1–12. *Unanswerable Objections to a Broadway Underground Railway* (New York: George F. Nesbitt & Company, 1873): 8–15.

In my judgment none of the other technical impediments was insurmountable. For instance, contractors had already cut through the hard Manhattan schist in completing two major public works during the 1840s: the digging of the Croton Aqueduct trench down Amsterdam Avenue and the blasting of a path down Park Avenue for the New York & Harlem Railroad. Technology was not the decisive factor in stopping the early subway proposals.

It should be noted that scholars disagree about this question. David C. Hammack asserted that until the mid-1890s technology—steam power's low speeds and poor ventilation and the lack of space for an open cut in Manhattan—"placed the most severe limits on rapid transit planning." David C. Hammack, *Power and Society: Greater New York at the Turn of the Century* (New York: Russell Sage Foundation, 1982): 237. James Blaine Walker argued that a subway was possible before electrification and believed that "not [steam] power but politics and the war for franchise rights" postponed underground construction. James Blaine Walker, *Fifty Years of Rapid Transit, 1864–1917* (New York: Law Printing Company, 1918): 15.

39. While the proposed New York subways would have been entirely dependent on the private sector, London's Metropolitan Railway owed its creation and its favorable balance sheet in large part to an unusual combination of financial subsidies. It had received substantial direct subsidy from the Corporation of the City of London and from the Great Western Railroad prior to construction. In addition, the Metropolitan also benefited from an indirect subsidy of increased traffic, as it was designed to allow Londoners to go to the passenger terminals of the Great Western, Great Northern, and Midland railroads. Barker and Robbins, *A History of London Transport,* v. 1, *The Nineteenth Century:* 99–113.

40. *New York Times,* February 11, 1912. Bobrick, *Labyrinths of Iron:* 192–94.

41. "Rapid Transit in New York," *Appleton's Journal:* 401.

42. Joseph Cunningham and Leonard O. DeHart, *A History of the New York Subway System,* Part 1, *The Manhattan Els and the IRT* (New York: n.p., 1976): 5. *New York Times,* July 4, 1868; March 12, 1898.

43. John Anderson Miller, *Fares, Please! From Horse-cars to Streamliners*

(New York: D. Appleton-Century Company, 1941): 71–73. Bobrick, *Labyrinths of Iron:* 197–200.

44. *New York Tribune,* July 11, 1885; *National Cyclopaedia of American Biography,* v. 11 (New York: James T. White and Company, 1909): 388–89. Miller, *Fares, Please!:* 75. Cunningham and DeHart, *A History of the New York Subway System,* Part 1, *The Manhattan Els and the IRT:* 5–9.

45. *New York Times,* May 13, 19, 20, 22; and July 1, 1875. New York State, *Laws of the State of New York, 1875* (Albany: Hugh J. Hastings, 1875): 740–56. Another restriction on government was imposed by the law's requirement that before the franchise could be assigned the commissioners had to secure the consents of the owners of more than half of the property, in value, along the proposed line, or else obtain a go-ahead from a commission appointed by the general term of the state supreme court.

46. New York City Board of Commissioners of Rapid Transit, *Proceedings, July to December, 1875* (New York: Martin B. Brown, 1877): 157–79. Cunningham and DeHart, *A History of the New York City Subway System,* Part 1, *The Manhattan Els and the IRT:* 9. For the evolution of Brooklyn's elevated lines, see James C. Greller and Edward B. Watson, *The Brooklyn Elevated* (Hicksville, NY: N.J. International, 1987?): 4–21.

47. William Fullerton Reeves, *The First Elevated Railroads in Manhattan and the Bronx of the City of New York* (New York: New-York Historical Society, 1936): 28–34.

48. The Manhattan Elevated Railway adopted a standard five-cent fare in 1886. For an account of elevated fares and collection methods, see Cunningham and DeHart, *A History of the New York Subway System,* Part 1, *The Manhattan Els and the IRT:* 14.

49. Iza Duffus Hardy, *Between Two Oceans; Or, Sketches of American Travel* (London: Hurst and Blackett, 1884): 99. See also *New York Herald,* August 1, 1880.

50. Hardy, *Between Two Oceans:* 99–100.

51. Ibid.: 101–2. In reporting that her trip took "little more than half an hour," Hardy probably employed literary license and slightly underestimated her actual traveling time.

52. James Francis Hogan, *The Australian in London and America* (London: Ward & Downey, 1889): 23.

53. William G. Marshall, *Through America; Or, Nine Months in the United States* (London: Sampson Low, Marston, Searle & Rivington, 1881): 24, 26–28.

54. William Dean Howells, *Impressions and Experiences* (New York: Harper and Brothers, 1896): 258–59.

55. Ibid.: 96, 98–99; G. W. Bromley and Company, *Atlas of the City of New York: Manhattan, 1898,* v. 3 (Philadelphia: G. W. Bromley and Company, 1898): plates 6, 7, 10, 11, 14.

56. NYSPSC, *Report, 1919,* v. 2: 12–17. U.S. Department of the Interior, Census Office, *Report on Transportation Business in the United States at the Eleventh Census: 1890,* Part 1, *Transportation by Land* (Washington, D.C.: Government Printing Office, 1895): 683–85.

Chapter 2. Making Government Safe for Business

1. SD, v. 9, entry for January 31, 1891, S&SP. In 1890 the state legislature passed a rapid transit act providing for a temporary rapid transit commission to report to the mayor on a practicable route. Mayor Hugh J. Grant appointed five commissioners: William Steinway, August Belmont, John H. Starin, Woodbury Langdon, and Orlando B. Potter. This commission failed to make a complete report due to a time limit contained in the act. This failure led to the passage of the Rapid Transit Act of 1891, creating a permanent commission. Despite this change from a temporary to a permanent commission, the Steinway Commission remained in the Husted Act tradition. NYCRTC, *Proceedings,* v. 2, 1899–1901 (New York; n.p., 1902): 767–68.

2. David C. Hammack, *Power and Society: Greater New York at the Turn of the Century* (New York: Russell Sage Foundation, 1982): 235–41. Charles W. Cheape, *Moving the Masses: Urban Public Transit in New York, Boston, and Philadelphia, 1880–1912* (Cambridge, MA: Harvard University Press, 1980): 74–80. New York State, *Laws of the State of New York, 1891* (Albany: Banks and Brothers, 1891): 3–19.

3. Undated statement of Jacob H. Schiff, Steinway & Sons Autograph Collection, Steinway Hall, New York, NY. This six-page statement signed by Schiff was addressed to "Mr. Chairman and Messrs. Commissioners." I believe Schiff drafted this statement in preparation for his testimony at a hearing of the Steinway commission on March 13, 1891. For a discussion of Schiff's testimony that day, see James Blaine Walker, *Fifty Years of Rapid Transit, 1864–1917* (New York: Law Printing Company, 1918): 133. In this statement Schiff argued that the creation of a commission dominated by the commercial elite should be a precondition of municipal funding, thus anticipating the link between business control and municipal investment that would be embodied in the Rapid Transit Act of 1894. For a profile of Schiff, see *National Cyclopaedia of American Biography,* v. 13 (New York: James T. White and Company, 1906): 533.

4. SD, v. 9, entry for February 16, 1891, S&SP. National Cyclopaedia of American Biography, v. 2 (New York: James T. White and Company, 1921): 514–15. *New York Times,* December 15, 1892. Ronald V. Ratcliffe, *Steinway & Sons* (San Francisco: Chronicle Books, 1989): 15–19, 28–30. Jon A. Peterson, ed., and Vincent F. Seyfried, cons., *A Research Guide to the History of the Borough of Queens and Its Neighborhoods* (New York: Queens College Department of History, 1983): 9. Vincent F. Seyfried, *300 Years of Long Island City, 1630–1930* (New York: Edgian Press, 1984): 69–74. I am grateful to Richard K. Lieberman for generously providing me with information about William Steinway.

5. New York City Board of Rapid Transit Railroad Commissioners, *Report to the Common Council of the City of New York in Pursuance of the Provisions of Section 5 of Chapter 4 of the Laws of 1891, October 20, 1891* (New York: n.p., 1891): 9–12. SD, v. 9, entries for February 9, October 20 and 28, 1891, and April 7 and 15, 1892, S&SP.

6. *New York Times,* December 15, 1892.

7. SD, v. 9, entries for December 6 and 13, 1892. Amory submitted two bids, one for $1,000 cash and the other for $500 cash and one-half percent of the gross receipts. SD, v. 9, entry for December 29, 1892, S&SP.

8. SD, v. 9, entries for December 30, 1892, January 2, 3, and 13, 1893, S&SP. *New York Times,* June 11 and 13, 1893. *Real Estate Record and Builders Guide,* January 7, 1893: 5–6.

9. SD, v. 9, entry for January 14, 1893, S&SP. Steinway ultimately agreed to support an elevated railway.

10. Julius Grodinsky, *Jay Gould: His Business Career, 1867–1892* (Philadelphia: University of Pennsylvania Press, 1957): 288–314. Maury Klein, *The Life and Legend of Jay Gould* (Baltimore: Johns Hopkins University Press, 1986): 282–91.

11. SD, v. 9, entries for March 24 and 29, April 12 and 18, May 5, July 14, and November 27, 1893, S&SP. *New York Times,* August 19, 1893.

12. *New York Times,* August 19, 1893.

13. *The National Cyclopaedia of American Biography,* v. 23 (New York: James T. White and Company, 1933): 360. *New York Times,* December 30, 1929.

14. *New York Times,* January 30 and February 1, 1894. SD, v. 10, entries for January 18, 19, 23, and 26, and February 1, 1894, S&SP. Chamber of Commerce of the State of New York, *Thirty-sixth Annual Report* (New York: Press of the Chamber of Commerce): 84–85.

15. Chamber of Commerce, *Thirty-sixth Annual Report:* 89.

16. *Ibid.:* 84–89.

17. *National Cyclopaedia of American Biography,* v. 13 (New York: James T. White and Company, 1906): 71–72. *New York Times,* June 4, 1914.

18. Chamber of Commerce, *Thirty-sixth Annual Report:* 91.

19. Ibid.: 95–102.

20. Ibid.: 112–25. *New York Evening Post,* April 23, 24, and 25, 1894; *New York Tribune,* April 26, 1894.

21. *New York Times,* April 26, 1894.

22. New York State, *Laws of the State of New York, 1894,* v. 2 (Albany: J. B. Lyon and Company, 1894): 1873–74. The five commissioners named in the act were Alexander E. Orr, William H. Steinway, John H. Starin, John Claflin, and Seth Low. By the time the Rapid Transit Commission began meeting, Alexander E. Orr had become president of the Chamber of Commerce and thus would have had two seats on the commission. Orr resigned his individual seat, and the Board named John H. Inman to replace him. For an analysis of the conflict between ward politicians and aristocratic businessmen in late nineteenth-century U.S. cities, see Jon C. Teaford, *The Unheralded Triumph: City Government in America, 1870–1900* (Baltimore: Johns Hopkins University Press, 1984): 15–82.

23. *New York Times,* February 1 and May 23, 1894; April 26, 1929. *New York Evening Post,* April 24, 1894. *New York Tribune,* April 27, 1894; NYCRTC, *Proceedings,* v. 1, *1894–98* (New York: n.p., 1899?): 79.

24. NYCRTC, *Proceedings,* v. 1, *1894–98:* 1–2, 21–45, 129–60.

25. *New York Times,* March 20, 1895.

26. Ibid.: 126, 189–224. *New York Times,* March 20 and 23, 1895. *Real Estate Record and Builders Guide,* March 16, 1895: 411–12; March 23, 1895: 456.

27. New York State, *Laws of the State of New York: 1894,* v. 2: 1874–76, 1885. A similar review process had been included in every rapid transit statute since the Husted Act of 1875. See New York State, *Laws of the State of New York: 1891:* 4–7.

28. SD, v. 10, entries for October 16 and 22, November 20, and December 18, 1895, and January 18 and March 7, 1896, S&SP. *New York Times,* March 8, 1896.

29. NYCRTC, *Proceedings,* v. 1, *1894–98:* 367.

30. *New York Times,* May 23, 1896.

31. NYCRTC, *Proceedings,* v. 1, *1894–98:* 365–71, 375–79, 405. SD, v. 10, entries for May 25, June 4, 11, and 18, July 17, and August 6, 1896, S&SP. *New York Times,* May 23, 1896.

32. In confirming the report of its special panel, the appellate division stipulated that the franchisee must pay a bond of $15 million to cover the period of construction and the period of the leasehold. This sum was later reduced to $5 million. NYCRTC, *Proceedings,* v. 1, *1894–98:* 609.

33. NYCRTC, *Proceedings,* v. 2, *1899–1901* (New York: n.p., 1902): 863–67.

34. Ibid.: 874–77; *National Cyclopaedia of American Biography,* v. 5 (New York: James T. White and Company, 1907): 481–82.

35. August Belmont to Lord Rothschild, June 15, 1892, CL 1, BFP.

36. NYCRTC, *Contract for Construction and Operation of Rapid Transit Railroad* (New York: n.p., 1900): 14–37. August Belmont to Baron Leopold de Rothschild, February 10, 1900, CL 3, BFP.

37. Historians disagree about Belmont's name change. David Black states that the banker's entire family adopted the name Belmont after French troops conquered the Rhineland during the Napoleonic wars. David Black, *The King of Fifth Avenue: The Fortunes of August Belmont* (New York: Dial Press, 1981): 4–6. Irving Katz refers to no name change at all. Irving Katz, *August Belmont: A Political Biography* (New York: Columbia University Press, 1968): 5–7. I found Stephen Birmingham's explanation that Belmont adopted his new name in response to American anti-Semitism persuasive. Stephen Birmingham, *"Our Crowd": The Great Jewish Families of New York* (New York: Harper & Row, 1967): 24–26, 73–77.

38. Katz, *August Belmont:* 7. Black, *The King of Fifth Avenue:* 17–25. Birmingham, *"Our Crowd":* 24–26.

39. *National Cyclopaedia of American Biography,* v. 37 (New York: James T. White and Company, 1951): 25–26. *Who's Who in New York City and State,* 1st ed. (New York: L. R. Hamersly, 1904): 49–50. *New York Times,* September 28, 1898; October 25, 1979. Eleanor Robson Belmont, *The Fabric of Memory* (New York: Farrar, Straus and Cudahy, 1957): 82–91, 99, 117.

Chapter 3. William Barclay Parsons and the Construction of the IRT

1. L. P. Gratacap, *The Geology of the City of New York,* 3rd ed. (New York: Henry Holt and Company, 1909): 30–44, 147–51. Christopher J. Schu-

berth, *The Geology of New York City and Environs* (Garden City, NY: Natural History Press, 1968): 1–9, 61–91.

2. Arthur Goodrich, "William Barclay Parsons," *World's Work* 4 (May 1903): 3467–71. *National Cyclopaedia of American Biography,* v. 14 (New York: James T. White and Company, 1910): 218. *New York Times,* May 10, 1932. Benson Bobrick, *Parsons Brinckerhoff: The First 100 Years* (New York: Van Nostrand Reinhold Company, 1985): 3–7. New York District Railway, *New York District Railway,* 9th ed. (New York: n.p., 1886): 4–7.

3. Goodrich, "William Barclay Parsons," *World's Work:* 3469–70.

4. "Memoir of William Barclay Parsons," *ASCE Proceedings* 59 (October 1933): 1485–92. William Barclay Parsons, "Rapid Transit in Great Cities," an address delivered March 13, 1905, at Purdue University: 1–2.

5. T. C. Barker and Michael Robbins, *A History of London Transport: Passenger Travel and the Development of the Metropolis,* v. 1, *The Nineteenth Century* (London: George Allen & Unwin, 1975): 296–97. There were three main technical problems: how to transmit electrical current from a stationary power station to a moving car; how to build a motor that could withstand repeated stoppings and startings and constant changes in speed; and where to mount the motor on the car so that it could withstand violent jolts and lurches. Harold C. Passer, *The Electrical Manufacturers: A Study in Competition, Entrepreneurship, Technical Change and Economic Growth* (Cambridge, MA: Harvard University Press, 1953): 213–18.

6. Passer, *The Electrical Manufacturers:* 217–55. U.S. Department of Commerce and Labor, Bureau of the Census, *Street and Electric Railways, 1902* (Washington, D.C.: Government Printing Office, 1903): 5.

7. William Barclay Parsons, *Report on Rapid Transit in Foreign Cities:* 1686–89. In New York State, *Assembly Documents,* 119th Session, 1896, v. 26, part 2 (Albany: Wynkoop Hallenbeck Crawford, 1896). SD, v. 9, entry for April 14, 1894, S&SP. Passer, *The Electrical Manufacturers:* 231, 241. Parsons, "Rapid Transit in Great Cities": 6–8.

8. Parsons included data from Berlin, Baltimore, and Chicago in his report, although he did not visit those cities on his trip. Parsons, *Report on Rapid Transit in Foreign Cities:* 1623, 1653–57, 1667–68.

9. Parsons, *Report on Rapid Transit in Foreign Cities:* 1629–39, 1681–89. Barker and Robbins, *A History of London Transport,* v. 1, *The Nineteenth Century:* 305–15. O. S. Nock, *Underground Railways of the World* (London: Adam and Charles Black, 1973): 35–39, 43–52. Karl Baedeker, *London and Its Environs,* 9th rev. ed. (Leipzig: Karl Baedeker, 1894): 113.

10. Nock, *Underground Railways of the World:* 44–45; B. H. M. Hewitt and S. Johannesson, *Shield and Compressed Air Tunneling* (New York: McGraw-Hill, 1922): 251–55. John O. Bickel and T. R. Kuessel, eds. *Tunnel Engineering Handbook* (New York: Van Nostrand Reinhold, 1982): 93–105.

11. Gratacap, *The Geology of the City of New York:* 30–44. Egbert L. Viele, *Report on the Typography and Hydrology of New York, to the Sanitary Association of the City of New York, June 13, 1859* (New York: Edmund Jones and Company, 1859): 3–6. Schuberth, *The Geology of New York City and Environs:* 61–91. The contour of the schist was not the only reason for the construc-

tion of skyscrapers in lower Manhattan and midtown; Wall Street had already been established as a business district, and the openings of Grand Central Terminal and Pennsylvania Station drew businesses to midtown.

12. Bickel and Koessel, eds., *Tunnel Engineering Handbook:* 103–5. Schuberth, *The Geology of New York City and Environs:* 1–9, 61–86. New York City Board of Rapid Transit Commissioners, *Report* (New York: n.p., 1891): 25–33. Parsons originally wanted to relocate the utility lines in pipe galleries built along the subway tracks. These galleries would have given repairmen direct access to the lines, ending the need to tear up the streets all the time. This idea was dropped, however, following the 1895 supreme court ruling that the subway was over the $50 million construction limit. David C. Hammack, *Power and Society: Greater New York at the Turn of the Century* (New York: Russell Sage Foundation, 1982): 252.

13. Parsons lost one major design decision about whether to build the express and local tracks on one level (as Chief Engineer William E. Worthen of the Steinway Commission wanted, and both the Steinway Commission and the Rapid Transit Commission ruled) or on top of one another (as Parsons preferred). See Charles S. Scott, "Design and Construction of the IRT: Civil Engineering," in Historic American Engineering Record, *Interborough Rapid Transit Subway (Original Line),* HAER NY-122 (Washington, D.C.: Historic American Engineering Record, 1978): 231–32. See also *Engineering News* 32 (December 27, 1894): 537–38.

14. Passer, *The Electrical Manufacturers:* 271–74. William Barclay Parsons to Edward M. Shepard, October 15, 1898, Edward M. Shepard Papers, Rare Book and Manuscript Library, Butler Library, Columbia University, New York, NY. Edward M. Shepard to William Barclay Parsons, January 21, 1899, Shepard Papers. William Barclay Parsons to Edward M. Shepard, February 26, 1899, Shepard Papers.

15. William Barclay Parsons, *An American Engineer in China* (New York: McClure, Philips and Company, 1900): 151–53. Bobrick, *Parsons Brinckerhoff:* 10–16.

16. *New York Times,* May 5 and June 17, 1901. Scott, "Design and Construction of the IRT: Civil Engineering": 263–68. Interview with Rosemary P. Pepe on November 27, 1989, New York, NY, LaGuardia and Wagner Archives, LaGuardia Community College/CUNY, Long Island City, NY.

17. *Engineering News* 47 (April 17, 1902): 318–20. Interborough Rapid Transit Company, *The New York Subway: Its Construction and Equipment* (New York: n.p., 1904): 37–39. Scott, "Design and Construction of the IRT": 237. Parsons, "Rapid Transit in Great Cities": 10–12.

18. Scott, "Design and Construction of the IRT": 244–48, 257–58. *Engineering News* 48 (September 18, 1902): 202. Interborough Rapid Transit Company, *The New York Subway:* 57–61.

19. *New York Times,* January 29, 1902.

20. Diary of William Barclay Parsons, Chief Engineer, Rapid Transit Commission, from the beginning of work, March 26, 1900, to his resignation as chief engineer, December 31, 1904, v. 2, entry for June 17, 1902, Rare Book and Manuscript Library, Butler Library, Columbia University, New York, NY. *New*

York Times, June 18 and 30, 1902. *Engineering News,* 48 (September 18, 1902): 202. NYCRTC, *Report, 1902* (New York: n.p., 1903): 223–24, 238–41.

21. *New York Times,* April 28, 1901. Interborough Rapid Transit Company, *The New York Subway:* 58–61. NYCRTC, *Report 1902:* 259–69. NYCRTC, *Report, 1903* (New York: n.p., 1904): 152–54.

22. NYCRTC, *Report 1903:* 153–54. Parsons, Diary, v. 2, entries for October 25 and 26, 1903. *New York Times,* October 25 and 26, 1903.

23. NYCRTC, *Report, 1903:* 153.

24. NYCRTC, *Report, 1903:* 153–54. Parsons, Diary, v. 2, entries for October 25 and 26, 1903.

25. Scott, "Design and Construction of the IRT": 267.

Chapter 4. The Subway and the City

1. *New York Times,* October 28, 1904. *New York Evening Post,* October 27, 1904.

2. August Belmont to William Barclay Parsons, July 17, 1904, PL 79, BFP.

3. *New York Times,* October 28, 1904.

4. Ibid.

5. London County Council, *The Rapid Transit Subways of New York: A Report by Mr. J. Allen Baker, Chairman of the Highways Committee, of the Inspection made by him of the Rapid Transit Subways of New York* (London: Southwood, Smith & Company, 1904): 5.

6. Interborough Rapid Transit Company, *The New York Subway, Its Construction and Equipment* (New York: n.p., 1904): 15, 23–26.

7. *New York Times,* October 28, 1904. *New York Mail & Express,* October 28, 1904. *Brooklyn Eagle,* October 27, 1904. Elmer Rice, "Joy-Riding in the Subway," *New Yorker* 4 (December 29, 1928): 21–23.

8. *New York Times,* October 28, 1904. *New York Evening Post,* October 27, 1904. Although the capacity of the original subway was six hundred thousand, only part of the route was opened on October 27.

9. *New York Times,* October 30 and 31, 1904. *New York Commercial,* October 31, 1904. *New York Tribune,* October 29, 1904. Abraham Lincoln Merritt, "Ten Years of the Subway," in Electric Railroaders Association, *Interborough Bulletin* 77 (November 1987): 12. The capacity of 350,000 was for the portion of the subway that was opened on October 27, 1904, not the entire IRT.

10. *Real Estate Record and Builders Guide,* November 5, 1904: 949. *New York Times,* November 19, 1904.

11. *New York Times,* October 29 and November 2, 5, 11, 18, 19, and 29, 1904; June 17, 1990.

12. Rice, "Joy-Riding in the Subway," 22, 23.

13. *Rand McNally & Co.'s Handy Guide to New York City,* 18th ed. (Chicago: Rand McNally & Company, 1905): 23. *Banner Guide & Excursion Book* (New York: John D. Hall, 1904): 4.

14. *Utica Saturday Globe,* November 5, 1904.

15. *New York Commercial,* October 31, 1904. *New York Times,* October 29 and 31, and November 2, 1904.

16. The IRT express service was so innovative that the subway's planners did not foresee its significance and originally assumed that the locals would carry the greater part of the passenger load, hence their decision to restrict express operations to built-up areas south of Ninety-sixth Street. Original specifications had called for building only two local tracks above Ninety-sixth Street, but a third track for one-way express service was added after construction began, partly in response to the traffic increases caused by electrification of the elevateds. Diary of William Barclay Parsons, Chief Engineer, Rapid Transit Commission, from the beginning of work, March 26, 1900, to his resignation as Chief Engineer, December 31, 1904," entry for January 28, 1901, Rare Book and Manuscript Library, Butler Library, Columbia University, New York, NY.

17. August Belmont to W. L. Bull, January 10, 1905, PL 81, BFP.

18. *New York Times,* March 23, 1905.

19. James O'Dean (lyrics) and Jerome Kern (music), "The Subway Express" (New York: T. B. Harms Company, 1907).

20. John F. Kasson, *Amusing the Million: Coney Island at the Turn of the Century,* American Century Series (New York: Hill & Wang, 1978): 41–50. Kathy Peiss, *Cheap Amusements: Working Women and Leisure in Turn-of-the-Century New York* (Philadelphia: Temple University Press, 1986): 121–22.

21. *New York Times,* November 4, 1904; February 12, 1908. Diary of Abraham Lincoln Merritt, entries for September 11, 1907, and August 5, 1910, Frank J. Sprague Library, Electric Railroaders Association, New York, NY. Merritt and the newspapers distinguished between accidental fatalities and suicides; see Merritt Diary entries for February 12 and December 26, 1907; February 19 and 20, 1908; and May 11, 1909. It should be noted that trolleys were capable of much faster speeds when unencumbered by traffic. My references to trolley speeds refer to streetcars that were operating in traffic. For an analysis of a similar phenomenon on the railroads, see John R. Stilgoe, *Metropolitan Corridor: Railroads and the American Scene* (New Haven, CT: Yale University Press, 1983): 167–88. See also Rosalind Williams, *Notes on the Underground: An Essay on Technology, Society, and the Imagination* (Cambridge, MA: MIT Press, 1992): 1–21, and Wolfgang Schivelbusch, *The Railway Journey: The Industrialization of Time and Space in the l9th Century* (Berkeley, CA: University of California Press, 1986): passim.

22. *New York Evening Post,* October 27, 1904.

23. Jill Stone, *Times Square: A Pictorial History* (New York: Collier Books, 1982): 23, 33–40.

24. Ibid.: 50–55. Freemont Rider, *Rider's New York* (New York: Henry Holt and Company, 1916): 7, 171. Meyer Berger, *The Story of the New York Times, 1851–1951* (New York: Simon & Schuster, 1951): 138, 144, 155–56.

25. Quoted in *Bartlett's Familiar Quotations,* 15th ed. (Boston: Little, Brown and Company, 1980): 759.

26. *Real Estate Record and Builders Guide,* June 16, 1906: 1137.

27. Mary C. Henderson, *The City and the Theatre: New York Playhouses*

from Bowling Green to Times Square (Clifton, NJ: James T. White and Company, 1973): 208–9. Stone, *Times Square:* 81–85, 94. Robert Sklar, *Movie-Made America: A Social History of American Movies* (New York: Random House, 1975): 4, 13–14, 45, 58. Richard Alleman, *The Movie Lover's Guide to New York* (New York: Harper & Row, 1988): 71–72, 90–91.

28. *Phillips' Business Directory of New York City, 1900–1901* (New York: W. Phillips, 1900): 1161–62. *Phillips' Business Directory of New York City, 1915* (New York: John F. White, 1915): 1150–51. *New York Times,* October 15, 1907. Henderson, *The City and the Theatre:* 186–209. Stone, *Times Square:* 34, 42–55, 60, 85–90, 102–14: Lewis A. Erenberg, *Steppin' Out: New York Nightlife and the Transformation of American Culture, 1890–1930* (Chicago: University of Chicago Press, 1981): 40–56.

29. Merritt, Diary, entries for October 29 and December 31, 1905; June 16, April 15 and 29, and September 29, 1907.

30. Walter Mack with Peter Buckley, *No Time Lost: The Autobiography of Walter Mack* (New York: Atheneum, 1982): 10.

31. C. T. Hill, "The Growth of the Upper West Side of New York," *Harper's Weekly* 40 (July 25, 1896): 730.

32. Ibid.: 730–34. Peter Salwen, *Upper West Side Story: A History and Guide* (New York: Abbeville Press, 1989): 47–75. New York City Department of Buildings, *Record of New Buildings, 1892–93:* 154. Building Docket Books, Municipal Archives, New York City Department of Records and Information Services, New York, NY.

33. James Trager, *West of Fifth: The Rise and Fall of Manhattan's West Side* (New York: Atheneum, 1987): 22–23. G. W. Bromley and Company, *Atlas of the City of New York: Manhattan, 1898,* v. 3 (Philadelphia: G. W. Bromley and Company, 1898): plates 6, 7, 10, and 14.

34. *Real Estate Record and Builders Guide,* June 10, 1911: 1091.

35. Lewis Mumford, *Sketches from Life* (New York: Dial Press, 1982): 4.

36. Ibid.: 6. See also Donald L. Miller, *Lewis Mumford: A Life* (New York: Weidenfeld & Nicholson, 1989): 3–33.

37. Edwin H. Spengler, *Land Values in New York in Relation to Transit Facilities* (New York: Columbia University Press, 1930): 145–62.

38. *New York Times,* September 19, 1909. *Real Estate Record and Builders Guide,* September 23, 1911: 415–16; March 27, 1915: 525; April 26, 1916: 650. U.S. Works Progress Administration, Federal Writers Project, *New York City Guide,* American Guide Series, rev. ed. (New York: Random House, c. 1939; reprint ed., New York: Random House, 1982): 284–86. Trager, *West of Fifth:* 60–64. W. A. Swanberg, *Citizen Hearst: A Biography of William Randolph Hearst* (New York: Charles Scribner's Sons, 1961): 255.

39. There was a station at Ninety-first Street that closed in 1959.

40. *Real Estate Record and Builders Guide,* November 7, 1911: 873; June 10, 1911: 1901. G. W. Bromley and Company, *Atlas of the City of New York: Borough of Manhattan, 1920–1921,* v. 3 (Philadelphia: G. W. Bromley and Company, 1920): plate 11. John Taurnac, *Essential New York: A Guide to the History and Architecture of Manhattan's Important Buildings, Parks, and Bridges* (New York: Holt, Rinehart and Winston, 1979): 127–28. Andrew

Alpern, *New York's Fabulous Luxury Apartments* (New York: Dover Publications, 1975): 50.

41. Ye Olde Settlers' Association of Ye West Side, *History, By-Laws, List of Members, and Historical Papers* (New York: n.p., 1921): 105.

42. Bronx Museum of the Arts, *Building a Borough: Architecture and Planning in the Bronx, 1890–1940* (New York: n.p., 1986): unpaginated brochure. G. W. Bromley and Company, *Atlas of the City of New York, Borough of the Bronx,* v. 2 (Philadelphia: G. W. Bromley and Company, 1913): plate 35. The photograph was taken at the intersection of Rosedale and Story avenues. Largely because this intersection was located about a mile from the IRT station and was situated on low-lying, marshy ground, its development took longer than expected and had a very different form than originally anticipated. After World War II this area became the site of three giant public housing projects—Bronxdale (1957), James Monroe (1961), and Rosedale (1961)—that effaced the original building lots and obliterated some of the streets that had been opened. One block north of Story Avenue, Ludlow Boulevard evolved into the busy Bruckner Expressway. E. Belcher Hyde, *Atlas of the Borough of the Bronx,* v. 4 (New York: E. Belcher Hyde, 1927, corrected to 1969): plates 44 and 45.

43. G. W. Bromley and Company, *Atlas of the City of New York, Borough of the Bronx* (Philadelphia: G. W. Bromley and Company, 1904): plates 23, 26, and 35.

44. Lloyd Ultan and Gary Hermalyn, *The Bronx in the Innocent Years, 1890–1925* (New York: Harper & Row, 1985): 152.

45. Due to the presence of land speculators, the spatial pattern that unfolded in the Bronx diverged from the leading historical model of the city-building process. According to Sam B. Warner's study of the Boston streetcar suburbs of Roxbury, West Roxbury, and Dorchester, the Massachusetts metropolis was "the product of thousands of separate decisions" made by large institutions such as street railways and waterworks as well as small operators such as speculative builders—all acting under common disciplines such as money markets, topography, and architectural styles. But major land speculators who were such a force in New York were not present in the Boston suburbs, which were old if sparsely settled towns where the land had been divided into smaller parcels that made the assemblage of large tracts difficult. New York's big speculators confined themselves to the undeveloped and unbroken land north of 125th Street; although Charles T. Barney acquired Upper West Side parcels such as the four vacant Broadway corners around the Eighty-sixth Street IRT stop, most of his transactions were made in northern Manhattan and the Bronx. Warner's model is more applicable to the Upper West Side. Sam B. Warner, *Streetcar Suburbs: The Process of Growth in Boston, 1870–1900* (Cambridge, MA: Harvard University Press, 1960; reprint ed., New York: Atheneum, 1968): 3–5. For accounts that follow Warner's lead, see James Leslie Davis, *The Elevated System and the Growth of Northern Chicago* (Evanston, IL: Northwestern University Department of Geography, 1965); Joel A. Tarr, "Transportation Patterns in Pittsburgh, 1850–1934," *Essays in Public Works History* 6 (April 1978); and Peter G. Goheen, *Victorian Toronto, 1850–1900: Pattern and Process of Growth* (Chicago: University of Chicago Department of Geography, 1970).

For an analysis of transport's impact on an existing urban area, see Olivier Zunz, "Technology and Society in an Urban Environment: The Case of the Third Avenue Elevated Railway," *Journal of Interdisciplinary History* 3 (Summer 1972): 89–102.

46. *National Cyclopaedia of American Biography,* v. 14 (New York: James T. White and Company, 1910): 409. *New York Times,* November 15, 1907. *Real Estate Record and Builders Guide,* February 27, 1904: 434; November 16, 1907: 799.

47. New York City Tenement House Department, *Fourth Report* (New York: n.p., 1909): 83–84. For the role of the IRT in stimulating a speculative boom in Harlem that contributed to the creation of a black ghetto there, see Gilbert Osofsky, *Harlem: The Making of a Ghetto,* 2nd ed. (New York: Harper & Row, 1963): 87–91.

48. New York City Tenement House Department, *Report for the Years 1912, 1913, and 1914* (New York: n.p., 1915): 121. Roy Lubove, *The Progressives and the Slums: Tenement House Reform in New York City, 1890–1917* (Pittsburgh: University of Pittsburgh Press, 1962): 111. In 1925 over 90 percent of New Yorkers lived within half a mile of transit facilities. Cities Census Committee, Inc., *Population of the City of New York, 1890–1930* (New York: n.p., 1932): 83–84. *New York World,* September 24, 1905.

49. Robert Coit Chapman, *The Standard of Living among Workingmen's Families in New York City* (New York: Charities Publication Committee, 1909): 75–84; G. W. Stevenson, "The Bronx and Beyond" in *The Wayfarer in New York,* ed. Edward S. Martin (New York: Macmillan, 1909): 231–34.

50. Cities Census Committee, Inc., *Population of the City of New York, 1890–1930:* 109.

51. Real Estate Board of New York, *Apartment Building Construction in Manhattan, 1902–1953* (New York: n.p., 1953): 8–11, 18–19.

Chapter 5. Good-bye to the Patricians

1. *New York Tribune,* October 28, 1904. *Real Estate Record and Builders Guide,* November 5, 1904: 949.

2. Cities Census Committee, Inc., *Population of the City of New York, 1890–1930* (New York: n.p., 1932): 51–53, 83–84. "New Subways for New York," *Outlook* 103 (March 15, 1913): 571–72.

3. *North Side News,* February 18, 1906.

4. NYSTC, *Ninth Annual Report: 1929* (Albany: J. B. Lyon and Company, 1930): 86–87. NYSPC, *Annual Report, 1910,* v. 3 (Albany: J. B. Lyon and Company, 1912): 25–31. Bion J. Arnold, *Reports upon the Interborough,* Report no. 6, *The Traffic of the Subway of the Interborough Rapid Transit Company of New York City* (New York: M. B. Brown, 1908): 15–17. Bion J. Arnold, *Reports upon the Interborough Subway,* Report no. 4, *The Capacity of the Subway of the Interborough Rapid Transit Company of New York City* (New York: M. B. Brown, 1908): 7. *New York Tribune,* October 7, 1905. Daniel L. Turner, "Is There a Vicious Circle of Transit Development and City Congestion?" *National*

Municipal Review 15 (June 1929): 322. Subway traffic also varied according to monthly and daily cycles. See Arnold, *Reports upon the Interborough Subway,* Report no. 6, *The Traffic of the Subway:* 9–14. For the daily cycle on a cable car, see Stephen Crane, "In the Broadway Cars" in *Last Words* (London: Digby, Long Company, 1902): 173–80.

5. Arnold, *Reports upon the Interborough Subway,* Report no. 4, *The Capacity of the Subway:* 7. Chicago Public Library, Municipal Reference Library, *A Study of Rapid Transit in Seven Cities,* Municipal Reference Library Bulletin No. 3 (Chicago: n.p., 1914): 14. U.S. Department of Commerce, Bureau of the Census, *Central Electric and Power Stations and Street and Electric Railway, 1912* (Washington, D.C.: U.S. Government Printing Office, 1915): 214–15.

6. Odette Keun, *I Think Aloud in America* (London: Longmans, Green and Company, 1939): 49–51. Haruko Ichikawa, *Japanese Lady in America* (Tokyo: Kenkyusha Company, 1938): 44. Chang Yee, *The Silent Traveler in New York* (New York: The John Day Company, 1950?): 32–41. Although the Tokyo subway is extremely crowded, it is significant perhaps that the subway employs guards to push people into the cars rather than forcing the riders to do so themselves. Cooper-Hewitt Museum, *Subways* (New York: Cooper-Hewitt Museum, 1977): 5.

7. During construction the IRT's concrete and steel framework was covered with up to eight layers of asphalt and felt waterproofing that retained the heat in the stations during the summer. George A. Soper, "The Condition of Air of the Rapid Transit Subway," *Street Railway Journal* 27 (March 31, 1906): 496. Interborough Rapid Transit Company, *The New York Subway: Its Construction and Equipment* (New York: n.p., 1904): 37–39. *The New Yorker* 3 (January 7, 1928): 10.

8. *New York Times,* March 9, 1914.

9. *New York Times,* July 4, 1910.

10. *New York Times,* February 24, 1912.

11. Hildegarde Hawthorne, *New York* (London: Adam and Charles Black, 1911): 40.

12. "The Discomforts of New York," *Outlook* 85 (January 5, 1907): 15.

13. Interborough Rapid Transit Company, *The New York Subway:* 125–34. *New York Times,* November 30, 1909.

14. Subway ridership was probably more weighted toward the middle class before 1920 than afterward due to the impact of the dual contract system in dispersing residents from working-class neighborhoods in Manhattan to the outer boroughs. This idea is developed in chapters 6 and 8.

15. Although the IRT formed a special police unit (the forerunner of the N.Y.C. Transit Authority police) to deal with the problems of rush hour congestion, these men generally did not face more serious criminal matters; the term "safety" referred to technical issues such as derailments and car design, and violent crime did not become a major concern until 1960–65. Kate Simon, *New York Places and Pleasures* (New York: Harper & Row, 1971): 9. Langston Hughes, "The Subway Rush Hour," in *On City Streets: An Anthology of Poetry,* ed. Nancy Larrick (New York: M. Evans and Company, 1968): 63. Joyce

Kilmer, *The Circus and Other Essays and Fugitive Pieces* (New York: George H. Doran Company, 1921): 91–92. Glen E. Holt, "The Changing Perception of Urban Pathology: An Essay on the Development of Mass Transit in the United States," in *Cities in American History,* ed. Kenneth T. Jackson and Stanley K. Schultz (New York: Alfred A. Knopf, 1972): 329.

16. "New York's Subway Problem: A Review," *Outlook* 101 (June 22, 1912): 386. See also *New York Times,* March 10, 1905.

17. *New York Times,* March 21, 1909.

18. *New York Times,* April 1, 1909.

19. *New York Times,* May 8, 1909.

20. *New York Times,* February 7, March 19, 27, and 29, April 1, July 1, and August 4, 1909. For an examination of the relationship between women and the automobile, see Virginia Scharff, *Taking the Wheel: Women and the Coming of the Automobile Age* (New York: Free Press, 1991): 165–75. The Hudson and Manhattan's decision to institute ladies' cars seems to have been partly motivated by its growing economic competition with the IRT; the H&M apparently wanted to show that it was more responsive to the public.

21. *New York Tribune,* April 4, 1905. *New York Times,* May 28, 1905. *New York World,* May 28, 1905.

22. S. D. V. Burr, "Projected Subway Lines in Greater New York," *Iron Age* 77 (February 1, 1906): 412.

23. NYCRTC, *Report 1905* (New York: n.p., 1906): 179–97. *New York Herald,* March 24, 1905. *New York Times,* March 31, 1905. *Brooklyn Standard-Union,* April 21, 1905.

24. *Brooklyn Times,* November 23, 1906. *New York Commercial Advertiser,* November 24, 1905. *New York World,* October 2, 1904. *New York Evening Telegram,* December 8, 1905.

25. *New York Sun,* March 23, 1905. *New York Tribune,* March 29, 1905. *New York World,* March 24, 1905.

26. *New York World,* March 29, 1905.

27. August Belmont to Frank Hedley, November 12, 1908, PL 102, BFP. August Belmont to Frank Hedley, March 8, 1911, PL 117, BFP. August Belmont to Alexander E. Orr, April 21, 1905, PL 82, BFP. August Belmont to Florence Nostrand, April 19, 1905, PL 82, BFP. *New York Times,* July 17, 1955. *Who's Who in New York,* 7th ed. (New York: Who's Who Publications, 1918): 501–2. Abraham Lincoln Merritt, Diary, entries for April 21 and September 11, 1905; May 11 and November 3, 12, 1907, Frank J. Sprague Library, Electric Railroaders Association, New York, NY. *New York Times,* April 22, 1905, April 30, 1950.

28. August Belmont to Alexander E. Orr, November 8, 1904, PL 80, BFP.

29. August Belmont to Lord Rothschild, January 27, 1905, CL 4a, BFP. These earnings included the IRT elevated division.

30. August Belmont to Louis G. Kaufman, March 9, 1916, PL 148, BFP.

31. August Belmont to Lord Rothschild, May 24, 1904, CL 4a, BFP. August Belmont to Alexander E. Orr, November 8, 1904, PL 80, BFP. August Belmont to Lord Rothschild, August 21, 1907, CL 4a, BFP. August Belmont to William A. Gaston, January 19, 1913, PL 130, BFP. *New York Evening Post,* August 7, 1907.

32. August Belmont to Lord Rothschild, November 28, 1902, CL 3, BFP. *New York Times,* December 27, 1905; January 27, 1906. *Real Estate Record and Builders Guide,* December 30, 1905: 1042. Cynthia M. Latta, "The Return on the Investment in the Interborough Rapid Transit Company" (Ph.D. dissertation, Columbia University, 1974): 67–82. The Metropolitan and Manhattan acquisitions hurt the Interborough in the long run. After the Metropolitan deal, the IRT discovered that the street railway company was grossly overcapitalized. More troubling was the Interborough's 1902 lease of the Manhattan Railway, which required that a fixed, guaranteed annual rental of 7 percent on the elevated's value of $60 million be paid to the owners. This contract became burdensome when the transit industry entered an inflationary period after World War I, and it siphoned off income from the subways to bondholders who did not invest anything in the system. From 1930 to 1939 Manhattan investors pocketed $28 million, a sum close to the subway division's profit of $30 million for the same interval. August Belmont to Louis G. Kaufman, March 9, 1916, PL 148, BFP. NYSPSC, *Annual Report, 1919,* v. 2 (Albany: J. B. Lyon and Company, 1920): 13–17. Harry J. Carman, *Street Surface Railway Franchises of New York City* (New York: Columbia University Press, 1919): 215–19. Under the terms of an agreement made in May 1922, the Manhattan rental was reduced to 3 percent in 1922 and then raised to 4 percent in 1923, before increasing to 5 percent. NYSTC, "Memorandum on Reorganization of Interborough Rapid Transit Company," December 9, 1924, Transit folder, box 33, George McAneny Papers, Seely G. Mudd Manuscript Library, Princeton University, Princeton, NJ. Interborough Rapid Transit Company, *Annual Reports,* 1927–39.

33. *New York Tribune,* December 23, 1905; *New York American,* December 24, 1905.

34. *New York Evening Post,* December 23, 1905.

35. New York State, *Laws of the State of New York, 1894,* v. 2 (Albany: J. B. Lyon and Company, 1894): 1873–99.

36. Wallace F. Sayre and Herbert Kaufman, *Governing New York City: Politics in the Metropolis* (New York: W. W. Norton, 1965): xxxv–30. For works examining such elite business domination of cities, see Roy Lubove, *Twentieth-Century Pittsburgh: Government and Environmental Change* (New York: John Wiley and Sons, 1969): vii–viii, 20–40, and Joe R. Feagin, *Free Enterprise City: Houston in Political and Economic Perspective* (New Brunswick, NJ: Rutgers University Press, 1988): 131–41.

37. August Belmont to Lord Rothschild, January 4, 1907, CL 4a, BFP; Ray Stannard Baker, "The Subway 'Deal,' " *McClure's* 24 (March 1905): 458, 462–63.

38. My discussion follows James B. Crooks' conclusion about the identity of the progressives. James B. Crooks, *Politics and Progressives: The Rise of Urban Progressivism in Baltimore, 1895–1911* (Baton Rouge, LA: Louisiana State University Press, 1968): 195–207. It also relies on Richard Hofstadter's view of the reform movement as a widespread "effort to restore a type of economic individualism and political democracy that was widely believed to have existed earlier in America and to have been destroyed by the great corporation and the corrupt political machine; and with that restoration to bring back a kind of morality and civic purity that was also believed to have been lost."

Richard Hofstadter, *The Age of Reform: From Bryan to F.D.R.* (New York: Vintage Books, 1955): 5–6. The main alternative interpretations include Samuel P. Hays' emphasis on modernization and efficiency, "The Politics of Reform in Municipal Government in the Progressive Era," *Pacific Northwest Quarterly* 55 (October 1964): 157–69; Gabriel Kolko's thesis that progressivism was a conservative attempt to preserve existing social and power relationships in a new economic setting, *The Triumph of Conservatism: A Reinterpretation of American History, 1900–1916* (Glencoe, IL: Free Press, 1963; reprint ed., Chicago: Quadrangle Books, 1967): 3; and Robert H. Wiebe's theory of the new middle class seeking to fulfill its destiny by rationalizing American society through so-called bureaucratic means, *The Search for Order, 1877–1900* (New York: Hill & Wang, 1967): 166.

39. Jacob A. Riis, *How the Other Half Lives: Studies Among the Tenements of New York* (New York: n.p., 1890; reprint ed., New York: Hill & Wang, 1957): 5–15. Kenneth T. Jackson, "Social Science Theory and New York City: Geographical and Historical Forces That Have Shaped the Built Environment," a paper presented at the Social Science Research Council Conference on the Landscape of Modernity, New York, NY, March 31–April 1, 1990: 6. Roy Lubove, *The Progressives and the Slums: Tenement House Reform in New York City, 1890–1917* (Pittsburgh: University of Pittsburgh Press, 1961): 111. Adna F. Weber, "Rapid Transit and the Housing Problem," *Municipal Affairs* 6 (Fall 1902): 408–17. H. L. Cargill, "Small Houses for Workingmen," in *The Tenement House Problem,* ed. Robert W. DeForest and Lawrence Veiller, v. 1 (New York: Macmillan, 1903): 331–35, 346–53. John DeWitt Warner, "Municipal Ownership Needed to Correlate Local Franchises," *Municipal Affairs* 6 (Winter 1902–3): 516. In reality these social critics did not understand poor ethnic families very well. They assumed that immigrant families continued to reside in the slums mainly because of the male breadwinner's need to reside close to his job. They missed women and children's contributions to household income through piecework, boarding and after-school jobs; they also failed to grasp the appeal to immigrants of familiar neighborhood and ethnic ties.

40. Baker, "Subway 'Deal,' " *McClure's:* 463.

41. Ibid.: 463. Warner, "Municipal Ownership Needed," *Municipal Affairs:* 516.

42. Warner, "Municipal Ownership Needed," *Municipal Affairs:* 516.

43. Baker, "Subway 'Deal,' " *McClure's*: 494; R. Fulton Cutting, "Shall New York Own Its Own Subway?" *Outlook* 79 (April 15, 1905): 931–33. For an analysis of a much different group of progressives—in San Francisco—who united with working-class representatives and championed municipal ownership, see Mark Ciabattari, "Urban Liberals, Politics, and the Fight for Public Transit, San Francisco, 1897–1915" (Ph.D. dissertation, New York University, 1988): vi–xii.

44. *New York Times,* March 11, 1905.

45. *New York Herald,* November 25, 1905.

46. "New York Rapid Transit," *Outlook* 82 (April 21, 1906): 866–67. New York State, *Laws of the State of New York, 1906,* v. 2 (Albany: J. B. Lyon and Company, 1906): 1103–20. *New York World,* February 16, 1906. *North Side News,* March 11, 1906. *New York Herald,* February 14, 1906.

47. *New York Times,* September 27 and October 23, 1906. W. A. Swanberg, *Citizen Hearst: A Biography of William Randolph Hearst* (New York: Charles Scribner's Sons, 1961): 239–52.

48. *New York Times,* March 28, April 5, May 6, 7, and 15, 1907. In a letter to the Rothschilds, however, Belmont claimed that the IRT did not oppose the bill because it would replace the incompetent, overly political state railroad commission and because Hughes could be trusted to appoint honest men. The purpose of this letter seems to have been to reassure the Rothschilds that regulatory change would not damage their investments. August Belmont to Lord Rothschild, January 4, 1907, CL 4a, BFP.

49. *New York Times,* March 28, May 16 and 23, and June 7, 1907. Robert F. Wesser, *Charles Evans Hughes: Politics and Reform in New York, 1905–1910* (Ithaca, NY: Cornell University Press, 1967): 153–69. New York State, *Laws of the State of New York, 1907,* v. 1 (Albany: J. B. Lyon and Company, 1907): 889–912.

Chapter 6. The Dual Contracts

1. Kenneth T. Jackson, "The Capital of Capitalism: The New York Metropolitan Region, 1890–1940," in *Metropolis, 1890–1940,* ed. Anthony Sutcliffe (Chicago: University of Chicago Press, 1984): 326. Ira Rosenwaike, *The Population History of New York City* (Syracuse, NY: Syracuse University Press, 1972): 133. Moses Rischin, *The Promised City: New York's Jews, 1870–1914* (Cambridge, MA: Harvard University Press, 1962): 76–90.

2. For an analysis of rapid transit as a solution to the problem of congestion, see Peter Derrick, "The Dual System of Rapid Transit: The Role of Politics and City Planning in the Second Stage of Subway Construction in New York City, 1902 to 1913" (Ph.D. dissertation, New York University, 1979): 210–64.

3. The Fourth Avenue route would also have a branch extending from Greenwood Cemetery to Coney Island. NYSPSC, *Annual Report, 1907,* v. 1 (Albany: J. B. Lyon and Company, 1908): 8–13. NYSPSC, *Proceedings, from July 1st to December 31st, 1907,* (New York: n.p., 1908): 414, 852, 925–30. NYSPSC, *Annual Report, 1910,* v. 1 (Albany: J. B. Lyon and Company, 1911): 49. NYSPSC, *Annual Report, 1912,* v. 1 (Albany: J. B. Lyon and Company, 1913): 59–61. *New York Times,* November 13, 1909. Milo R. Maltbie, "The Fruits of Public Regulation in New York," *Annals of the American Academy of Political and Social Science* 48 (January 1911): 177–78.

4. The five public service commissioners included the chairman, William R. Willcox, a Republican party stalwart who had served as commissioner of Manhattan parks during the administration of Mayor Seth Low and who was an old friend of Governor Charles Evans Hughes; Edward M. Bassett, an ex-Democratic congressman from Brooklyn who knew Governor Hughes from their days at Columbia Law School and who helped draft the bill that established the Public Service Commission; Milo R. Maltbie, a former editor of the highly regarded reform journal *Municipal Affairs* and an economist who became an expert on public utilities regulation; John E. Eustis, a lawyer and former Bronx parks commissioner; and retired Brooklyn businessman William McCarroll. *The*

National Cyclopaedia of American Biography, v. A (New York: James T. White and Company, 1930): 445. *Who's Who in New York,* 8th ed. (New York: Who's Who Publications, Inc., 1924): 418, 825. *Who's Who in New York,* 7th ed. (New York: Who's Who Publications, 1918): 1154.

5. *New York World,* January 2, 1908.

6. NYCRTC, *Report 1905* (New York: n.p., 1906): 179–97. William R. Willcox, "The Transportation Problem in New York City," *Harper's Weekly* 54 (March 5, 1910): 13, 30. NYSPSC, *Annual Report, 1907,* v. 1: 10–13, 33–35. *New York Evening Post,* January 2, 1908. The People's Institute, Forum Committee, *A Communication to the Board of Estimate and Apportionment in Favor of a Competitive Subway System* (New York: n.p., 1910?): 3.

7. The Brooklyn Rapid Transit Company had one direct connection between Brooklyn and Manhattan, a spur line that ran over the Brooklyn Bridge and carried a heavy volume of traffic. James C. Greller and Edward B. Watson, *The Brooklyn Elevated* (Hicksville, NY: N.J. International, Inc., n.d.): 15.

8. *National Cyclopaedia of American Biography,* v. 14 (New York: James T. White and Company, 1910): 382. *National Cyclopaedia of American Biography,* v. 44 (New York: James T. White and Company, 1962): 548–49. *New York Times,* February 6, 1910. Edward M. Bassett, *The Autobiography of Edward M. Bassett* (New York: The Harbor Press, 1939): 107–15, 118–23. Willcox, "Transportation Problem": 13. Milo R. Maltbie, "Rapid Transit Subways in Metropolitan Cities," *Municipal Affairs* 4 (September 1900): 458–80. Stanley Buder, "Edward M. Bassett," *Dictionary of American Biography,* Supplement 4, ed. John A. Garraty and Edward T. James (New York: Charles Scribner's Sons, 1974): 55–57. NYSPSC, *Annual Report, 1907,* v. 1: 34–35. Edward M. Bassett, "Distribution of Population in Cities," *American City* 13 (July 1915): 7–8. Edward M. Bassett, "Municipal Ownership of Public Utilities," *American City* 8 (April 1913): 364–66.

9. "The New York Campaign: The Real Owner," *Outlook* 93 (October 30, 1909): 488.

10. *New York Times,* May 7, 1909, and October 25, 1910. "New York's Subway Policy," *Charities and Commons* 21 (March 13, 1909): 1201–3. Carl W. Condit, *The Port of New York: A History of the Rail and Terminal System from the Beginnings to Pennsylvania Station* (Chicago: University of Chicago Press, 1980): 263–93.

11. The other members of the committee were bankers Paul M. Warburg, Howard C. Smith, Clarence H. Kelsey, and J. Edgar Leaycroft.

12. Julius Henry Cohen, *They Builded Better Than They Knew* (New York: Julian Messner, 1946): 271–88. *Who's Who in New York (City and State) 1924,* 8th ed. (New York: Who's Who Publications, Inc., 1924): 964. Chamber of Commerce of the State of New York, *Fifty-first Annual Report* (New York: Press of the Chamber of Commerce, 1909): 151–55, 162–67. Chamber of Commerce of the State of New York, *Fifty-second Annual Report* (New York: Press of the Chamber of Commerce, 1910): 60–66.

13. Chamber of Commerce of the State of New York, *Fifty-third Annual Report* (New York: Press of the Chamber of Commerce, 1911): 120–21.

14. Lincoln Steffens, *The Shame of the Cities* (New York: McClure, Phil-

lips and Company, 1904; reprint ed., New York: Hill & Wang, 1957): 1–18, 206–14. Ida M. Tarbell, *The History of the Standard Oil Company* (New York: Macmillan Company, 1904; reprint ed., Gloucester, MA: Peter Smith, 1963): vii–xiii.

15. *New York Times,* January 18, 1909. NYSPSC, *Annual Report, 1908,* v. 1 (Albany: J. B. Lyon and Company, 1909): 28.

16. Milo R. Maltbie, *Report on the Indeterminate Franchise for Public Utilities* (New York: n.p., 1908): 4. Two other proposals were designed to improve the municipality's capacity to fund subway building. One proposal involved a new system of underwriting rapid transit that would replace the old methods of appropriating tax revenues and issuing city bonds with a new method, assessing the land that adjoined the new routes so that property holders who profited from the subway's impact on rising land values would pay for the improvements. The other measure involved the elimination of a significant barrier to rapid transit construction: the constitutional requirement that municipal bonds for subway construction be included within the city's debt limit of 10 percent of the assessed valuation of real estate. Arguing that subways, unlike roads, bridges, schools, and most other capital projects, were revenue-producing, the Public Service Commission favored an amendment to the state constitution to exclude rapid transit bonds from the debt limit. *New York Times,* March 3, 1909. Chamber of Commerce, *Fifty-first Annual Report:* 3–5. NYSPSC, *Annual Report, 1908,* v. 1 (Albany: J. B. Lyon and Company, 1910): 17–23.

17. The PSC's assessment proposal was also approved by May 1909. Its constitutional debt amendment was not authorized until the following November, however. Enactment of an amendment to the New York State constitution required passage by separate sessions of the legislature and approval by the voters in a referendum; in the spring of 1909 the legislature approved the proposal to exempt self-sustaining subway (and dock) bonds from the debt limit for the second time and sent it to the voters, who endorsed it that November. *New York Times,* January 7 and 18, April 7 and 24, May 27, and November 14, 1909; Derrick, "Dual System": 182–83. New York State, *Laws of the State of New York,* v. 2 (Albany: J. B. Lyon and Company, 1909): 1200–70.

18. Wallace S. Sayre and Herbert Kaufman, *Governing New York City: Politics in the Metropolis* (New York: Russell Sage Foundation, 1960): 626. In 1909 the Board of Estimate comprised New York's three citywide officials (the mayor, comptroller, and president of the Board of Aldermen, who had three votes apiece) and the presidents of its five boroughs (the presidents of the most populous boroughs, Manhattan and Brooklyn, cast two votes each, and the presidents of the three smaller boroughs, the Bronx, Queens, and Richmond, had one vote apiece). Seymour J. Mandelbaum, *Boss Tweed's New York* (New York: John Wiley & Sons, 1965): 106. David C. Hammack, *Power and Society: Greater New York at the Turn of the Century* (New York: Russell Sage Foundation, 1982): 225–28.

19. Both of these powers were acquired from the Board of Aldermen. Derrick, "The Dual System": 25, 97.

20. Ibid.: 258–60.

21. William Hochman, "William J. Gaynor: The Years of Fruition" (Ph.D.

dissertation, Columbia University, 1955): 108. *National Cyclopaedia of American Biography,* v. 18 (New York: James T. White and Company, 1922): 289–90. Lately Thomas, *The Man Who Mastered New York: The Life and Opinion of William J. Gaynor* (New York: William Morrow and Company, 1969): 327–35. Edwin R. Lewinson, "John Purroy Mitchel, Symbol of Reform" (Ph.D. dissertation, Columbia University, 1961): 71–78. *New York Times,* September 24, October 9, 20, 29 and 30, and November 4, 1909.

22. Hochman, "William J. Gaynor": 387–93. *National Cyclopaedia of American Biography,* v. 16 (New York: James T. White and Company, 1918): 353–54.William J. Gaynor, "The Looting of New York," *Pearson's Magazine* 21 (May 1909): 461–73. Mortimer Smith, *William J. Gaynor, Mayor of New York City* (Chicago: Henry Regnery Company, 1951): 21–35, 58–72. Lewis Heaton Pink, *The Tammany Mayor Who Swallowed the Tiger* (New York: International Press, 1931): 196–202.

23. Willcox, "The Transportation Problem in New York City": 13.

24. Derrick, "The Dual System": 265–388; *New York Times,* February 17, October 21, 25, and 28, 1910.

25. Smith, *William Jay Gaynor:* vii, ix. Although Gaynor's latest about-face startled started New Yorkers who had followed his campaign denunciations of the transit companies, it was actually comprehensible. A nineteenth-century economic conservative who was skeptical of active government and who feared municipal extravagance, Gaynor concluded that expanding the Interborough would be more prudent than constructing Triborough. Ibid.: 122. Pink, *The Tammany Mayor Who Swallowed the Tiger:* 196–201. Derrick, "The Dual System": 315–24. Hochman, "William J. Gaynor": 398–420.

26. *New York Times,* August 23, 1910. Citizens' Committee [of the Chamber of Commerce and the Merchants' Association], *The Rapid Transit Problem of New York City* (New York: n.p., 1910): 1–29.

27. Chamber of Commerce, *Fifty-first Annual Report:* 157–67.

28. NYSPSC, *Proceedings,* v. 5, *From January 3 to December 31, 1910* (New York: n.p., 1911): 583–84, 697, 706–7. NYSPSC, *Annual Report, 1910,* v. 1 (Albany: J. B. Lyon and Company, 1911): 33–34, 42–44.

29. *New York Times,* November 20, 1910.

30. Burton J. Hendrick, "McAdoo and the Subway," *McClure's* 36 (March 1911): 468. William Gibbs McAdoo, *Crowded Years: Reminiscences of William G. McAdoo* (Boston: Houghton Mifflin Company, 1931): 1, 8, 31, 36, 348. *National Cyclopedia of American Biography,* v. 62 (Clifton, NJ: James T. White and Company, 1984): 131–33.

31. Robert Watchorn, "The Builder of the Hudson Tunnels," *Outlook* 90 (December 1909): 909–14. Hendrick, "McAdoo and the Subway": 485–500. McAdoo, *Crowded Years:* 65–97. The Hudson and Manhattan completed the northern tunnel in September 1905, the downtown tunnel in December 1907, the Thirty-third Street extension in 1910, and the Newark extension in 1912. Condit, *The Port of New York: A History of the Rail and Terminal System from the Beginnings to Pennsylvania Station:* 249–58, 385. Brian J. Cudahy, *Rails under the Mighty Hudson: The Story of the Hudson Tubes, the Pennsy Tunnels and Manhattan Transfer* (Brattleboro, VT: Stephen Greene Press, c. 1975): 6–9.

32. McAdoo estimated that the subway would cost $150 million. He offered to pay $50 million for equipment and proposed that the city contribute $100 million for the purchase of real estate and the cost of construction. The city would own the subway and lease it to the Hudson and Manhattan for a period of time determined during negotiations. The new subway would be operated separately from the Hudson tubes, and riders would have to pay an extra fare to transfer from one system to another. William Gibbs McAdoo to William R. Willcox, November 18, 1910. NYSPSC, *Proceedings,* v. 5, *From January 3 to December 31, 1910* (New York: n.p., 1911): 759–66.

33. *New York Times,* November 19 and 20, 1910; *New York Evening Post,* November 21, 1910.

34. August Belmont to Lord Rothschild, December 2, 1910, CL 5, BFP.

35. NYSPSC, *Proceedings,* v. 5, *From January 3 to December 31, 1910:* 459–61, 494.

36. August Belmont to Gardiner M. Lane, May 24, 1909, PL 4a, BFP.

37. August Belmont to Lord Rothschild, December 2, 1905, CL 5, BFP.

38. August Belmont to Gardiner M. Lane, May 24, 1909, CL 4a, BFP.

39. August Belmont to Gardiner M. Lane, April 22, 1903, PL 73, BFP. August Belmont to Lord Rothschild, December 2, 1910, CL 5, BFP. *New York Times,* December 6, 1910. T. P. Shonts to William R. Willcox, December 5, 1910: 67–71, in NYSPSC, *Annual Report, 1910,* v. 1 (Albany: J. B. Lyon and Company, 1911). "A Great Rapid Transit System for a Great City," *Scientific American* 103 (December 17, 1910): 478. In addition to the subways and elevateds, Shonts proposed that the PSC approve the completion of the long-stalled Steinway Tunnel from Forty-second Street in Manhattan to Hunters Point in Queens. The IRT would pay all but $1.5 million of the $10 million needed for the Steinway Tunnel.

40. *New York Times,* December 4, 1910.

41. John J. Broaslame, "William Gibbs McAdoo: Businessman in Politics" (Ph.D. dissertation, Columbia University, 1970): 77.

42. William Gibbs McAdoo to William R. Willcox, December 15, 1910: 5–56, in NYSPSC, *Annual Report, 1910,* v. 1. Derrick, "The Dual System": 300, 483, 503.

43. *New York Evening Post,* December 6, 1910.

44. *New York Times,* December 6, 1910.

45. Ibid.; Derrick, "Dual System": 358.

46. Lewinson, "John Purroy Mitchel": 73–75. *New York Times,* December 21 and 31, 1910.

47. Joseph Cunningham and Leonard DeHart, *A History of the New York Subway System,* Part 2, *Rapid Transit in Brooklyn* (New York: n.p., 1977): 9–29. Greller and Watson, *The Brooklyn Elevated:* 3–29.

48. H. H. Porter to Edwin W. Winter, September 26, 1906, Edwin W. Winter Papers, Archives/Manuscripts Division, Minnesota Historical Society, St. Paul, MN. "Diary of William Barclay Parsons, chief engineer, Rapid Transit Commission, from the beginning of work, March 26, 1900, to his resignation as chief engineer, December 31, 1904," entries for April 26 and May 5, 1902, and February 9, 1903, Rare Book and Manuscript Library, Butler Library, Columbia

University. August Belmont to Seth Low, March 27, 1903, PL 72, BFP. *New York Tribune,* January 11, 1911. Edwin W. Winter to William R. Willcox, March 2, 1911: 423–27; Edwin W. Winter to William R. Willcox and George McAneny, April 25, 1911: 427–34; and Edwin W. Winter to George McAneny and William R. Willcox, May 2, 1911, in NYSPSC, *Annual Report, 1911,* v. 1: 434. In addition to the subways, the BRT also wanted to add third tracks to its existing elevateds and to run its trains over the Williamsburg Bridge. Under the terms of this BRT proposition, the City of New York would pay for the construction of the Fourth Avenue and bridge loop subways, while the BRT would cover the charges of converting its railroads into rapid transit facilities. The BRT wanted to have 6 percent of the existing value of the lines that would be tied to the Fourth Avenue and the costs of interest and sinking fund on the capital that it would borrow for the improvements. After these BRT expenses had been covered from operating revenues, the city would receive money to pay for the interest charges on its bonds; if any operating revenues remained after the city and the company had covered their expenses, then the city and the company would divide it equally.

49. August Belmont to Leopold de Rothschild, May 16, 1911, CL 5, BFP.

50. George McAneny, "What I Am Trying to Do," *World's Work* 26 (June 1913): 177.

51. Mel Scott, *American City Planning since 1890* (Berkeley, CA: University of California Press, 1970): 1–15. 110–82. Derrick, "The Dual System": 274–76. Paul Boyer, *Urban Masses and Moral Order in America, 1820–1920* (Cambridge, MA: Harvard University Press, 1978): 264–76. William H. Wilson, *The City Beautiful Movement in Kansas City* (Columbia, MO: University of Missouri Press, 1964): xiii–xvii, 120–38.

52. Alva Johnson, "Little Giant, Parts I and II," *The New Yorker* 6 (May 17 and 24, 1930): 29–32, 24–27. James Lardner, "Painting the Elephant," *The New Yorker* 60 (June 25, 1984): 41–72.

53. "George McAneny," *American* 74 (September 1912): 544.

54. *New York Times,* July 30, 1953. *Who's Who in America,* v. 19 (Chicago: A. N. Marquis Company, 1936): 1629. James Weinstein, *The Corporate Ideal in the Liberal State: 1900–1918* (Boston: Beacon Press, 1968): ix–xv. Although the comptroller is an elected office, McAneny was appointed to the position to fill a vacancy. Henry F. Griffin, "The Reformer in Office," *Outlook* 98 (August 12, 1911): 829–32. Henry F. Griffin, "A Keeper of the City," *Outlook* 93 (November 27, 1909): 659–61. McAneny, "What I Am Trying to Do": 172–81.

55. New York City Board of Estimate and Apportionment, *Minutes, January 1 to March 31, 1911* (New York: Lecouver Press, 1911): 279–81, 522. *New York Times,* January 20 and 28, 1911.

56. Derrick, "The Dual System": 386.

57. NYSPSC, *Proceedings 1911,* v. 6, *From January 3 to December 30, 1911* (New York: n.p., 1912): 122–24, 281–84.

58. New York City Board of Estimate and Apportionment, *Minutes, April 1 to June 30, 1911* (New York: Lecouver Press, 1911): 2367–430.

59. "George McAneny," *American:* 544.

60. August Belmont to Lord Rothschild, July 21, 1911, CL5, BFP.

61. "New York's Subway Problem: A New Offer," *Outlook* 100 (March 9, 1912): 517. "New York's Subway Problem: A Review," *Outlook* 101 (June 22, 1912): 384–88. *New York Times,* January 9 and 16, February 29, and May 25, 1912. NYSPSC, *Proceedings, 1912,* v. 7, *From January 2 to December 31, 1912* (New York: n.p., 1913): 436–41. *New York Times,* January 16 and 22, May 23, 24, and 25, 1912. In addition to the two contracts, the PSC also issued, as part of the agreements with the companies, certificates for the third-tracking and extension of various elevated lines of the Manhattan Railway Company and the Brooklyn Rapid Transit Company. Agreements were also made for joint trackage rights. See Derrick, "Dual System": 494–563.

62. NYSPSC, *Contract No. 3* (New York: J. W. Pratt, 1913): 65–77. NYSPSC, *Contract No. 4* (New York: J. W. Pratt, 1913): 59–70. Delos F. Wilcox, "The New York Subway Contracts," *National Municipal Review* 2 (July 1913): 375–91. LeRoy T. Harkness, "The Dual System Contracts in Their Relation to the Rapid Transit History of New York City," *Proceedings of the Municipal Engineers of the City of New York for 1913* (New York: n.p., 1914): 258–84.

63. Derrick, "Dual System": 594.

64. Boris S. Pushkarev with Jeffrey M. Zupan and Robert S. Cumella, *Urban Rail in America: An Exploration of Criteria for Fixed-Guideway Transit* (Bloomington, IN: Indiana University Press, 1982): table A-2. It should be noted that *Urban Rail in America* employs route miles, not track miles, to measure the length of rapid transit lines and thus does not take into account New York's extensive express system; had statistics for track miles been used, New York's lead over London and other cities would have been greater. *Urban Rail in America*'s statistics for New York are for the region, not for the city; accordingly, I have subtracted the Hudson and Manhattan's 22.2 route miles from the New York region's 1920 total. *Jane's Urban Transport Systems* (London: Jane's Publishing Company, 1984): 366.

65. NYSPSC, *Contract No. 3:* 14–23. NYSPSC, *Contract No. 4:* 12–19. Daniel L. Turner, "Brooklyn's Rapid Transit System," a paper read March 20, 1923, before the Kiwanis Club of Brooklyn: 4–6.

66. Cunningham and DeHart, *A History of the New York City Subway System:* Part 2, *Rapid Transit in Brooklyn* (New York: n.p., 1977): 36.

67. Isaac F. Marcosson, "The World's Greatest Traffic Problem," *Munsey's* 49 (June 1913): 338.

68. I am indebted to Edward Gray for his help with this formulation.

Chapter 7. Across the East River

1. In this chapter, Jackson Heights is examined as an example of a neighborhood that was shaped by the opening of the dual system. The decision to focus on Jackson Heights was made for the following reasons: It lay along the IRT subway route that ran through the Steinway tunnel; it was located far enough from Manhattan to demonstrate the dual contracts' impact on residential development in the outer boroughs, and it was a product of the Queensboro

Corporation, one of the highly capitalized and vertically organized enterprises beginning to dominate real estate development. Because it was a middle-class area and its apartments were tenant-owned rather than rented, Jackson Heights was not "typical" of all the neighborhoods spawned by the dual system. But Jackson Heights represents one pattern of development, a pattern of business city-planning that also shaped other New York City areas such as Rego Park and Kew Gardens, Queens.

2. The eight dual system crossings included the following: (1) the IRT Lexington Avenue's tunnel to the Bronx, (2) the BMT's Sixtieth Street tunnel, (3) the Steinway tunnel, (4) the BMT's Fourteenth Street tunnel, (5) the IRT's Clark Street, (6) the BMT's Montague Street tunnel, (7) the BMT's Williamsburg Bridge line, and (8) the BMT's Manhattan Bridge line. I have excluded the Queensboro Bridge line, which was used by elevated trains. Although the Williamsburg Bridge line was part of the Centre Street loop and opened before the dual contracts, it was operated as part of the dual system. The Queensboro line originally terminated at 103rd–104th Street (then Alburtis Avenue) in Corona. The extension to Flushing was completed on January 21, 1928. Joseph Cunningham and Leonard O. DeHart, *A History of the New York City Subway System,* Part 2, *Rapid Transit in Brooklyn* (New York: n.p., 1977): 30–35. Joseph Cunningham and Leonard O. DeHart, *A History of the New York City Subway System,* Part 1, *The Manhattan Els and the I.R.T.* (New York: n.p., 1976): 41–48, 60.

3. Carl W. Condit, *The Port of New York: A History of the Rail and Terminal System from the Beginnings to Pennsylvania Station* (Chicago: University of Chicago Press, 1980): 16–75. Brian J. Cudahy, *Rails under the Mighty Hudson: The Story of the Hudson Tubes, the Pennsy Tunnels and Manhattan Transfer* (Brattleboro, VT: Stephen Greene Press, c. 1975): 6–9.

4. "The New York and Long Island Tunnel," *Scientific American Supplement* 29 (June 21, 1890): 12057. *Prospectus of the New York & Long Island Railroad* (New York: n.p., 1895?): title page, 3–6, box 005, S&SP; "New York Tunnel Scheme," *Scientific American* 64 (January 24, 1891): 52. "The New York and Long Island Railroad Company Statement," July 1892, box 004, S&SP. Along with local traffic across the East River, the NY & LIRR was also expected to tap the through traffic to Montauk, at the tip of Long Island, and on to New England. Following the 1892 explosion, the New York and Long Island's scope was redesigned, apparently in an effort to attract English capital. It now included a tunnel under the Hudson River to provide access to the New Jersey railroads and a viaduct over the East River, at Hell Gate, to connect with the New England railroads. *Prospectus of the New York & Long Island Railroad Company:* 3–4, box 005, S&SP. New York and Long Island Railroad Company, "List of Stockholders representing the first $20,000.—," January 15, 1891, box 005, S&SP.

5. "Copy of a Certified Copy of Articles of Association of the New York and Long Island Railroad Company," 1891, box 005, S&SP. New York and Long Island Railroad Company, "Copy of Articles of Association, Dated July 25, 1887," box 004, S&SP. SD, v. 8, entry for May 11, 1887, S&SP. SD, v. 9, entries for December 10, 1890, July 9, 1891, and October 21, 1891, S&SP.

William Steinway later served as vice-president and as chairman of the NY & LIRR.

6. New York State, Office of the Secretary of State, "The People of the State of New York to the New York and L.I.R.R. Co.," January 5, 1891, box 004, S&SP. Long Island City Board of Aldermen, "Copy of Resolution Passed by the Board of Aldermen of Long Island City, granting right of way, etc.," n.d., box 005, S&SP. New York City Board of Aldermen, "Resolution of the Board of Aldermen for the Construction of a tunnel by the New York and Long Island Railroad Company," December 31, 1890, box 004, S&SP. New York and Long Island Railroad Company, "Abstract for Minutes of the Meeting of the Commissioner of the Land Office held at the Office of the Secretary of State on Tuesday, the twenty-fifth day of November, 1890," box 004, S&SP.

7. *Long Island Daily Star,* December 30, 1892; *New York Times,* December 29, 1892. Vincent F. Seyfried, *300 Years of Long Island City, 1630–1930* (New York: Edgian Press, 1984): 133–34.

8. Seyfried, *300 Years of Long Island City:* 133–34. *The New York and Long Island Railroad Company:* 3–19. Cudahy, *Rails under the Mighty Hudson:* 6–9. "New York Tunnel Schemes," *Scientific American* 64 (January 24, 1891): 52. Condit, *The Port of New York:* 249–51. H. B. Hammond to Walter C. Foster, August 9, 1892, box 004, S&SP. Walter C. Foster to Louis Von Bernuth, August 9, 1892, box 005, S&SP.

9. Vincent F. Seyfried, *The New York & Queens County Railway and the Steinway Lines, 1867–1939* (New York: n.p., 1950): 26–30. Vincent F. Seyfried, *The Story of the Long Island Electric Railway and the Jamaica Central Railways, 1894–1933* (New York: n.p., 1951): 1, 26. *Long Island Daily Star,* December 9, 1904, and May 5, 1905. *New York Times,* May 28, 1902, and September 25, 1907. *New York Tribune,* February 2, 1903; NYSPSC, *Documentary History of Railroad Companies* (Albany: n.p., 1914): 534, 783. For the sake of consistency I refer to the Steinway tunnel by that name throughout this chapter. It should be noted, however, that the Interborough called the tunnel the Belmont tunnel and the Public Service Commission named it the Queensboro subway.

10. *Long Island Daily Star,* May 5, July 14, and August 4, 1905. *New York Press,* July 11, 1905.

11. Arthur B. Reeve, "The Romance of Tunnel Building," *World's Work* 13 (December 1906): 8343.

12. *Long Island Daily Star,* September 29, 1905.

13. Reeve, "The Romance of Tunnel Building": 8338–51. *Long Island Daily Star,* August 4, 11, and 25, and September 29, 1905. Charles Scott, "The Design and Construction of the IRT: Civil Engineering," in Historic American Engineering Record, *The Interborough Rapid Transit Subway (Original Line),* HAER NY-122 (Washington, D.C.: Historic American Engineering Record, 1979): 262. Robert S. Mayor, "Shield Tunneling," in John O. Bickel and T. R. Kuessel, eds., *Tunnel Engineering Handbook* (New York: Van Nostrand Reinhold Company, 1982): 93–105.

14. *Long Island Daily Star,* August 18, 1905. Reeve, "The Romance of Tunnel Building": 8350. John Vipond Davies, "The Astoria Tunnel under the East River," *ASCE Proceedings* 41 (August 1915): 1405–85.

15. Reeve, "The Romance of Tunnel Building": 8338. S. D. V. Burr, "The Development of Tunneling in New York City," *Iron Age* 79 (January 10, 1927): 119–28.

16. *New York Times,* July 3, 1906.

17. Ibid. *New York Times,* November 15, 1916. "Another Tunnel Beneath the East River Completed," *Scientific American,* n.s. 97 (October 20, 1907): 286. Charles Prelini, "New York Subaqueous Tunnels—No. 1," *Engineering* 83 (November 8, 1907): 297–98. Göster E. Sandström, *Tunnels* (New York: Holt, Rinehart and Winston, 1962): 234–36. L. P. Gratacap, *Geology of the City of New York,* 3rd ed. (New York: Henry Holt and Company, 1909): 112–19. Reeve, "The Romance of Tunnel Building": 8338–42. Sandhogs who worked in compressed air were also subject to the bends, a painful and sometimes fatal condition that resulted when overly rapid reduction of atmospheric pressure caused the dissolution of nitrogen bubbles in the bloodstream. Although the cause of the bends had been discovered during the construction of the Brooklyn Bridge, the condition continued to threaten tunnel workers. David McCullough, *The Great Bridge: The Epic Story of the Building of the Brooklyn Bridge* (New York: Simon & Schuster, 1972): 281–322. "The New York & Long Island Tunnel," *Scientific American Supplement,* no. 75: 12057. Reeve, "The Romance of Tunnel Building": 8349. *Long Island Daily Star,* October 6, 1905. *New York Times,* November 24, 1905, and February 24, 1907.

18. *New York Times,* May 15 and 17, August 8 and 15, and September 24 and 25, 1907.

19. Jon A. Peterson, ed., and Vincent F. Seyfried, cons., *A Research Guide to the History of the Borough of Queens and Its Neighborhood* (New York: Queens College Department of History, 1983): 1–6. Frank Bergen Kelley, *Historical Guide to the City of New York* (New York: n.p., 1909): 277–93. Ira Rosenwaike, *Population History of New York City* (Syracuse, NY: Syracuse University Press, 1972): 58. Janet E. Lieberman and Richard K. Lieberman, *City Limits: A Social History of Queens* (Dubuque, IA: Kendall/Hunt Publishing Company, 1983): 66–70, 105–18.

20. G. W. Bromley and Company, *Atlas of the City of New York: Borough of Queens* (Philadelphia: G. W. Bromley and Company, 1909): plates 13 and 17. John H. Hendrickson, "After Field Plover," August 29, 1904, after field plover (and related materials) folder, JHHP. Henry Isham Halzelton, *The Boroughs of Brooklyn and Queens, Counties of Nassau and Suffolk, Long Island, New York, 1609–1924,* v. 2 (New York: Lewis Historical Publishing Company, 1925): 942. Jesse Lynch Williams, "Rural New York City," *Scribner's* 30 (August 1901): 178–91. Helen Fuller Orton, "Jackson Heights: Its History and Growth," a paper read on January 17, 1950, before the Newtown Historical Society, Newtown, Queens; on deposit at the Queens Borough Public Library, Jamaica, NY.

21. Kelley, *Historical Guide to the City of New York:* 279.

22. Diary of John H. Hendrickson, entry of September 15, 1900, JHHP. John H. Hendrickson, "Trip for Snipe, Friday morning, March 27, 1902," snipe hunting material folder, JHHP. Diary of John H. Hendrickson, entry for October 11, 1900, snipe hunting material folder, JHHP. Vincent F. Seyfried to Clifton Hood, May 20, 1986, in author's possession.

23. Diary of John H. Hendrickson, entry for September 14, 1900, JHHP.

24. Peter J. Schmitt, *Back to Nature: The Arcadian Myth in Urban America* (New York: Oxford University Press, 1969): xvii–xviii.

25. Diary of John H. Hendrickson, entry for September 14, 1900, untitled diary folder, JHHP.

26. John H. Hendrickson, "Typescript labeled 1909," untitled 1910? folder, JHHP.

27. *Brooklyn Eagle,* March 23, 1905. *New York Evening Post,* May 1, 1905. *Brooklyn Standard-Union,* July 6, 1905. *New York Times,* August 1, 1905; September 25, 1907. "Immediate Connection Between Steinway Tunnel and Grand Central Urged," *Queens Chamber of Commerce Bulletin* 2 (May 1915): 56. NYSPSC, *Contract No. 3* (New York: J. W. Pratt Company, 1913): 8, 19–21, 28–29.

28. Condit, *The Port of New York: A History of the Rail and Terminal System from the Beginnings to Pennsylvania Station:* 312–41. Norval White and Elliot Willensky, *AIA Guide to New York City,* rev. ed. (New York: Collier Books, 1978): 580–81. "Name New Tunnel 'Queensboro Subway,' " *Queensborough* 2 (August 1915): 67. Cunningham and DeHart, *A History of the New York City Subway System,* Part 1, *The Manhattan Els and the IRT:* 42–45. The dual contracts also authorized the extension of the IRT's Second Avenue elevated across the Queensboro Bridge to Bridge Plaza. Ibid.: 43–45.

29. *New York Times,* September 3, 1944. *National Cyclopaedia of American Biography,* v. 32 (New York: James T. White and Company, 1945): 458–59.

30. *Newtown Register,* January 20, 1910. Daniel Karatzas, *Jackson Heights: A Garden in the City* (New York: n.p., 1990): 10–15.

31. Karatzas, *Jackson Heights:* 18–19. Marian Stabler, "From Farm Lands Grew a Garden Community of Homes," *Fair Long Island* 2 (May 1938): 4–5. Alan F. Kornstein, "Jackson Heights: Biography of an Urban Community" (unpublished undergraduate paper, Yale University, 1972): 1–7, 12–13. For the dispute over which individual Jackson Heights was named for, see Karatzas, *Jackson Heights:* 18.

32. *Long Island Daily Star,* January 20, 1910; February 1, 1912; October 24, 1914; August 26 and December 2, 1915. Karatzas, *Jackson Heights:* 24–27.

33. Karatzas, *Jackson Heights:* 26.

34. *Souvenir Edition of the Jackson Heights News, Published upon the Opening of the New Executive Building of the Queensborough Corporation* (New York: n.p., 1929): 2. *Queensborough* 6 (October–November 1920): 407. *Queensborough* 7 (April 1921): 172. *Queensborough* 4 (May 1917): 65–75. For the original vision of Queens Boulevard as a grand tree-lined avenue similar to the Grand Concourse in the Bronx and Ocean Parkway and Eastern Parkway in Brooklyn, see Queens Borough Chamber of Commerce, *Queens Borough: Being a Descriptive and Illustrated Book of the Borough of Queens, City of Greater New York* (New York: n.p., 1913): 80–82, and Maurice E. Griest, "Ornamental Concrete Elevated Railway, New York City," *Engineering News* 74 (November 11, 1915): 913–18.

35. Ebenezer Howard, *Garden Cities of To-Morrow* (Cambridge, MA: MIT Press, 1965): 9, 41–47. Stanley Buder, *Visionaries and Planners: The Garden*

City Movement and the Modern Community (New York: Oxford University Press, 1990): 64–76, 157–80. Frederic J. Osborn, *Green-Belt Cities* (New York: Schocken, 1969): 40–52, 181–82. Jeffrey A. Kroessler and Nina S. Rappaport, *Historic Preservation in Queens* (New York: Queensborough Preservation League, 1990): 60, 64–65.

36. It should be noted that the Queensboro Corporation itself engaged in speculative activity prior to World War I, selling some of its vacant land to builders and some of its apartment buildings to investors. Following the development of the garden suburb plan, the corporation continued to sell property to builders; however, it enforced restrictions on building designs to ensure architectural cohesion and social homogeneity. Karatzas, *Jackson Heights:* 25–28, 154–59. For analyses of the city-building patterns in other suburbs, see William S. Worley, *J. C. Nichols and the Shaping of Kansas City: Innovation in Planned Residential Communities* (Columbia, MO: University of Missouri Press, 1990): 1–10, 21–36; Kenneth T. Jackson, *Crabgrass Frontier: The Suburbanization of the United States* (New York: Oxford University Press, 1985): 93–95, 123, 172–87; Robert Fishman, *Bourgeois Utopias: The Rise and Fall of Suburbia* (New York: Basic Books, 1987): 3–17, 116–33; and Mary Corbin Sies, "American Country House Architecture in Context: The Suburban Ideal of Living in the East and Midwest, 1877–1917" (Ph.D. dissertation, University of Michigan, 1987): 1–21. Marc A. Weiss, *The Rise of the Community Builders: The American Real Estate Industry and Urban Land Planning* (New York: Columbia University Press, 1987): 1–12; Michael H. Ebner, *Creating Chicago's North Shore: A Suburban History* (Chicago: University of Chicago Press, 1988): xvii–xxx. For Jackson Heights' place in the physical expansion of New York City during the interwar period, see Robert A. M. Stern, Gregory Gilmartin, and Thomas Mellins, *New York 1930: Architecture and Urbanism Between the Two World Wars* (New York: Rizzoli, 1987): 479–85.

37. Stabler, "From Farm Lands Grew a Garden Community of Homes": 5. "Jackson Heights," *Queensborough* 6 (December 1920): 435–37. *Souvenir Edition of the Jackson Heights News:* 3–4. Karatzas, *Jackson Heights:* 28–39. Jackson, *Crabgrass Frontier:* 4–11. Ann Durkin Keating, *Building Chicago: Suburban Developers and The Creation of a Divided Metropolis* (Columbus, OH: Ohio State University Press, 1988): 1–9.

38. *Souvenir Edition of the Jackson Heights News:* 16.

39. Several apartment complexes, including Ivy, Cedar, and Hayes Courts, were marketed as rental units rather than sold as cooperatives. Karatzas, *Jackson Heights:* 41–52. "Collective Ownership Plan," *Jackson Heights News,* n.d. 1919. Stabler, "From Farmlands Grew a Garden Community of Homes": 5. Jean Holmes, *Where to Live in New York* (New York: New York American, 1934): 54–57. The cooperative ownership plan was originally entitled the collective ownership plan. In addition to building apartments, the Queensboro also established meat markets, grocery stores, barbershops, drugstores, and other commercial businesses.

40. Karatzas, *Jackson Heights:* 54–69. John Taylor Boyd, Jr., "Garden Apartments in Cities," *Architectural Review* 48 (July 1920): 53–74. Andrew J. Thomas, "New Garden Apartments in Queens County, New York City,"

Architectural Forum 30 (June 1919): 187–91. Andrew J. Thomas, "The Evolution and Development of Jackson Heights," *Architectural Forum* 41 (August 1924): 61–66. Andrew J. Thomas, "The Button-Controlled Elevator in a New Type of Moderate-Price Apartment Buildings at Jackson Heights, New York City," *Architectural Record* 51 (May 1920): 486–90. The Queensboro Corporation did not develop all of the land itself; for the activities of other builders, see Karatzas, *Jackson Heights:* 78–80. Although the Queensboro Corporation concentrated on and became famous for its apartment buildings, the company also built single-family homes in Jackson Heights. See *Jackson Heights News,* August 21, 1925.

41. *Jackson Heights News,* December 20, 1919.

42. Margaret Marsh, *Suburban Lives* (New Brunswick, NJ: Rutgers University Press, 1990): xi–xvi, 92–103. *Jackson Heights News,* April 24, 1925.

43. *Jackson Heights News,* December 20, 1919.

44. *Jackson Heights News,* December 20, 1916; September 27, 1917; October 4 and December 1, 1922; and December 1, 1928. *Newtown Register,* February 6, 1919; April 17, 1919; June 26, 1919; October 16, 1919; and June 3, 1920. "Just as Important as Dumb Waiters," *Playground* 16 (December 122): 412–13. L. L. Little, "From a Sporting Center to a Center of Sport," *Outing* 78 (June 1921): 107, 135. "How the Community Idea Functions in Jackson Heights," *Playground* 19 (October 1925): 400–1.

45. Queensboro Corporation, *Jackson Heights* (New York: n.p., 1933): 4–5.

46. *Souvenir Edition of the Jackson Heights News:* 12, 16.

47. *Queensborough* 14 (June 1928): 343. *Queensborough* 7 (April 1921): 172. See also *Jackson Heights News,* March 27, 1936; Karatzas, *Jackson Heights:* 79; Worley, *J. C. Nichols and the Shaping of Kansas City:* 147–55; Marsh, *Suburban Lives:* 68–71. In 1927 the James Conforti Company, which owned several apartment buildings in Jackson Heights, challenged Queensboro's anti-Semitic housing discrimination in New York State Supreme Court. In an attempt to prevent a Queensboro subsidiary, the Jackson Heights Investing Company, from foreclosing on these apartments, the Conforti Company argued in an anti-Semitic thrust of its own that its failure to make its mortgage payments was due to the company's exclusion of Jews, which had limited the pool of potential tenants. Supreme Court Justice Burt J. Humphrey ruled that a private company such as the Jackson Heights Investing Company had the right to choose its tenants and could exclude anyone it wished. *New York Times,* February 2, 1927. Jackson Heights Investing Company v. James Conforti Construction Company, 222 A.D. 687 (1927). To my knowledge, discrimination against African-Americans did not become a public issue in Jackson Heights in the 1920s and 1930s, apparently because relatively few African Americans could afford housing there. Although McDougall gave Jackson Heights a Protestant cast in his promotional material, many Catholics settled there, and a Catholic parish, St. Joan of Arc, was organized in 1920. In 1919 the *Newtown Register* noted that the population was 60 percent Catholic; they seemed to be primarily Irish and German Catholics. *Newtown Register,* February 6, 1919.

48. *New York Times,* February 2, 1927. Jackson Heights Investing

Company v. James Conforti Construction Company, 222 A.D. 687 (1927). *Jackson Heights News,* March 27, 1936. Karatzas, *Jackson Heights:* 79. Worley, *J. C. Nichols and the Shaping of Kansas City:* 147–55. Marsh, *Suburban Lives:* 68–71.

49. *Souvenir Edition of the Jackson Heights News:* 2. Karatzas, *Jackson Heights:* 69–73, 117.

50. Karatzas, *Jackson Heights:* 116–34, 146–48, 154–62. It should be noted that the interior courtyards of the prewar apartment complexes remained intact and were not developed. It should also be noted that the Queensboro Corporation abandoned its prewar efforts to enforce aesthetic standards on buildings that were erected in the area. Also, residents of Jackson Heights apparently objected to particular decisions—such as the one not to build a park—but there seems to have been little organized resistance to these changes. Rather than fight, many of the original residents moved away, usually to Nassau and Westchester counties. The lack of more organized resistance suggests that Queensboro's efforts to build a community was futile after all. It should be noted that Jackson Heights experienced a demographic change in the 1950s when Jewish residents began to settle there. By the 1960s the area was widely perceived to be a Jewish section. Karatzas, *Jackson Heights:* 157–73. The Community Council of Greater New York, Bureau of Community Statistical Services, Research Division, *Queens Community Population Characteristics and Neighborhood Social Services,* v. 1 (New York: n.p., 1958): 38, 45–47, 58.

51. Rosenwaike, *Population History of New York City:* 133, 141. Cities Census Committee, Inc., *Population of the City of New York, 1890–1930* (New York: n.p., 1932): tables 15–20. Regional Plan of New York and Its Environs, *Regional Survey of New York and Its Environs,* v. 2, *Population, Land Values and Government* (New York: Regional Plan of New York and Its Environs, 1929): 76. Eleanora W. Schoenebaum, "Emerging Neighborhoods: The Development of Brooklyn's Fringe Areas, 1850–1930" (Ph.D. dissertation, Columbia University, 1976): 50–61, 78–80, 199–275. Deborah Dash Moore, *At Home in America: Second Generation New York Jews* (New York: Columbia University Press, 1981): 4–58. The population of the fifth and smallest borough, Richmond (Staten Island), increased 202 percent from 1910 to 1940. However, Staten Island remained isolated from the rest of the city and had not yet gone through its period of suburban development; in 1940 its 174,441 inhabitants accounted for only 2 percent of the total city population. Rosenwaike, *Population History of New York City:* 133.

52. Rosenwaike, *Population History of New York City:* 133. In 1960, Queens became the second most populous borough, pushing Manhattan into third place.

53. Ibid. These figures underestimate the reduction of Manhattan's population density since they do not take into account the expansion of northern Manhattan.

54. Donna Gabaccia, "Little Italy Tenements," a paper delivered at the Social Science Research Council Conference on the Landscape of Modernity, New York City, NY, December 1–2, 1990: 1–2, 15–21. Nancy L. Green, "Sweatshop Migrations: The Garment Industry Between Home and Shop,"

a paper delivered at the Social Science Research Council Conference on the Landscape of Modernity, New York City, NY, December 1–2, 1990: 14–21. Deborah Dash Moore, "Jewish Neighborhoods in Three Boroughs," a paper delivered at the Social Science Research Council Conference on the Landscape of Modernity, New York City, NY, December 1–2, 1990: 1–16. Mary Elizabeth Brown to Clifton Hood, March 5, 1992; in author's possession. Consolidated Edison Company of New York, Industrial and Economic Development Department, *Population Growth of New York City by Districts, 1910–1948* (New York: n.p., 1948): tables 1 and 2.

Chapter 8. John F. Hylan and the IND

1. U.S. Department of Commerce, Bureau of Economic Analysis, *Long Term Economic Growth, 1860–1970* (Washington, D.C.: U.S. Government Printing Office, 1973): 222–23. *New York Times,* July 25, 1918; January 19 and 20, 1919; and March 28, 1920. It should be noted that wage increases did not keep pace with inflation. See James J. McGinley, *Labor Relations in the New York Rapid Transit Systems, 1904–1944* (New York: King's Crown Press, 1949): 164–76.

2. August Belmont, Alexander E. Orr, Abram S. Hewitt, and William Barclay Parsons did not give much thought to the possibility that changes in the value of currency would affect subway revenues; if they had done so, they probably would have been more concerned about deflation—the chief monetary trend of the late nineteenth century—than about inflation. New York State, *Laws of the State of New York, 1894,* v. 1 (Albany: J. B. Lyon and Company, 1894): 1878. NYCRTC, *Contract for the Construction and Operation of Rapid Transit Railroad with Supplementary Agreements, February 21, 1900* (New York: n.p., 1900): 176. NYSPSC, *Contract No. 4* (New York: J. W. Pratt, 1913): 70–71. NYSPSC, *Contract No. 3* (New York: J. W. Pratt, 1913): 167. For early transit fares, see George Rogers Taylor, "The Beginnings of Mass Transportation in Urban America, Part II," *Smithsonian Journal of History* 1 (Fall 1966): 50. For the evolution of fares on New York's elevateds, see Joseph Cunningham and Leonard O. DeHart, *A History of the New York City Subway System,* Part 1, *The Manhattan Els and the IRT* (New York: n.p., 1976): 14. For the everyday value of the nickel in the 1890s and early 1900s, see Thomas J. Schlereth, *Victorian America: Transformations in Everyday Life* (New York: HarperCollins, 1991): 79–85. For conflicts over the fare in Los Angeles, see Robert C. Post, "The Fair Fare Fight: An Episode in Los Angeles," *Southern California Quarterly* 52 (September 1970): 275–98. For the English tradition of special working-class fares, see H. J. Dyos, "Workmen's Fares in South London, 1860–1914," *Journal of Transport History* 1 (May 1953): 3–19.

3. *New York Evening Post,* October 6–7, 1904. *New York Times,* October 6 and 16–18, 1919. U.S. Department of Commerce, *Long-Term Economic Growth, 1860–1970:* 222–23.

4. *New York Times,* January 5, 1918; December 14, 1919.

5. NYSPSC, *Report 1919,* v. 2 (Albany: n.p., 1920): 16–17. The BRT

charged a ten-cent fare to Coney Island. Joseph Cunningham and Leonard DeHart, *A History of the New York City Subway System,* Part 2, *Rapid Transit in Brooklyn* (New York: n.p., 1977): 24. See also Cynthia Latta, "The Return on the Investment of the Interborough Rapid Transit Subway" (Ph.D. dissertation, Columbia University, 1975): passim.

6. Interborough Rapid Transit Company, *1917–1918 Annual Report* (New York: John Ward & Son, 1918): 22–23. Interborough Rapid Transit Company, *1921–1922 Annual Report* (New York: John Ward & Son, 1922): 14–15. Brooklyn Rapid Transit Company, *Annual Report for the Year Ended June 30, 1917* (New York: n.p., 1917): unpaginated. Brooklyn-Manhattan Transit Corporation, *Annual Report 1925* (New York: n.p., 1925): 8.

7. For an analysis of the structural problems of excessive debt and inflated stocks and bonds that contributed to the railways' postwar financial crisis, see Glen E. Holt, "The Changing Perception of Urban Pathology: An Essay on the Development of Mass Transit in the United States," in *Cities in American History,* ed. Kenneth T. Jackson and Stanley K. Schultz (New York: Alfred A. Knopf, 1972): 333–38; Donald N. Dewess, "The Decline of the American Street Railways," *Traffic Quarterly* 24 (October 1970): 563–81; and Stanley Mallach, "Origins of the Decline of Urban Mass Transportation in the United States, 1890–1930," *Urbanism Past and Present* 8 (Summer 1979): 1–17.

8. Delos F. Wilcox, "The Transit Crisis in New York," *Nation* 108 (June 1919): 40.

9. "Sixty-one Fare Changes," *Electric Railway Journal* 52 (November 9, 1918): 856. "Thirty-one Fare Increases," *Electric Railway Journal* 52 (December 7, 1918): 1025. "The Traction Crisis in New York," *Review of Reviews* 61 (March 1920): 330. Harlow C. Clark, "Fare Increases in New York," *Electric Railway Journal* 55 (January 1920): 32–35.

10. James J. Flink, *America Adopts the Automobile, 1895–1910* (Cambridge, MA: MIT Press, 1970): 12–13, 23, 64–70, 203–31. John B. Rae, *The Road and the Car in American Life* (Cambridge, MA: MIT Press, 1971): 41–49. John B. Rae, *The American Automobile: A Brief History* (Chicago: University of Chicago Press, 1965): 5–17, 59, 87. Mark S. Foster, *From Streetcar to Superhighway: American City Planners and Urban Transportation, 1900–1940* (Philadelphia: Temple University Press, 1981): 59. U.S. Department of Transportation, Federal Highway Administration, *Highway Statistics Summary to 1975* (Washington, D.C.: U.S. Government Printing Office, 1975): 45.

11. Paul Barrett, *The Automobile and Urban Transit: The Formation of Public Policy in Chicago, 1900–1930* (Philadelphia: Temple University Press, 1983): xi–8. See also Foster, *From Streetcar to Superhighway:* 53–54; U.S. Congress, Senate, Committee on the Judiciary, Subcommittee on Antitrust and Monopoly, *American Ground Transportation: A Proposal for Restructuring the Automobile, Truck, Bus, and Rail Industries,* 93rd Congress, 2nd Session. (Washington, D.C.: U.S. Government Printing Office, 1974): 26–38; and Scott L. Bottles, *Los Angeles and the Automobile: The Making of the Modern City* (Berkeley, CA: University of California Press, 1987): 4–21. NYSPSC, *Report 1919,* v. 2 (Albany: n.p., 1920): 13–14. It should be noted that there were at least two points of competition between rapid transit and the automobile for street space: the presence of the els' pillars on the street contributed to the movement

to eliminate them, and the belief that subway kiosks interfered with street traffic led to their demolition.

12. Holt, "The Changing Perception of Urban Pathology": 333–38. Wilfred Owen, *The Metropolitan Transportation Problem,* rev. ed. (Garden City, NY: Doubleday, 1966): 26–65. Although the last major phase of subway construction was completed in 1940, New York did add some lines after that date. For the construction of the Chrystie Street extension and the Dyre Avenue line, the purchase of the Rockaway line from the Long Island Railroad, and the planning of the 63rd Street–Archer Avenue subway, see Joseph Cunningham and Leonard DeHart, *A History of the New York City Subway System,* Part 3, *The Independent System and City Ownership,* (New York: n.p., 1977): 33–34, 39, 44–46, 52, 54, 56. For a proposal to extend the Independent subway, which was aborted by the depression, see *New York Times,* September 16, 1929; American Public Transit Association, *1990 Transit Fact Book* (Washington, D.C.: American Public Transit Association, 1990): 11, 45, 52. For another city where transit remained important, see Joel A. Tarr, "Transportation Innovation and Changing Spatial Patterns in Pittsburgh, 1850–1934," *Essays in Public Works History* (April 1978): 24–28, 36; Metropolitan Transportation Authority, *A Methodology Report on the Comprehensive Travel Telephone Survey* (New York: n.p., 1990): 4.

13. *New York Times,* May 13 and 16, 1918; January 2, 1919.

14. John F. Hylan, *Autobiography of John Francis Hylan* (New York: Rotary Press, 1922): 14–24. *New York Times,* January 12, 1936. Joseph D. McGoldrick, "John Francis Hylan," *Dictionary of American Biography,* v. 22, ed. Edward Livingston Schuyler and Edward T. James (New York: Charles Scribner's Sons, 1958): 330–31.

15. Hylan, *Autobiography:* 24–25. *New York Times,* November 7, 1917.

16. Nancy Joan Weiss, *Charles Francis Murphy, 1858–1924: Respectability and Responsibility in Tammany Politics* (Northampton, MA: Smith College, 1968): 50–51. Niven Busch, Jr., "The Emerald Boss," *The New Yorker* 3 (March 12, 1927): 25–28. W. A. Swanberg, *Citizen Hearst: A Biography of William Randolph Hearst* (New York: Charles Scribner's Sons, 1961): 307–10. Interview with Robert F. Wagner, Jr., March 1, 1988, New York City, NY, LaGuardia and Wagner Archives, LaGuardia Community College/CUNY, Long Island City, NY. Peter E. Derrick, "The New York City Transit Crisis of 1918–1925" (master's essay, Columbia University, 1967): passim.

17. Kenneth S. Chern, "The Politics of Patriotism: War, Ethnicity, and the New York Mayoral Campaign," *New-York Historical Society Quarterly* 63 (October 1979): 291–313. The Republican candidate, William M. Bennett, received 8 percent of the vote. Ira Rosenwaike, *Population History of New York City* (Syracuse, NY: Syracuse University Press, 1972): 101. Edwin R. Lewinson, "John Purroy Mitchel, Symbol of Reform" (Ph.D. dissertation, Columbia University, 1961): passim.

18. William Bullock, "Hylan," *American Mercury* 1 (April 1924): 444.

19. McGoldrick, "John Francis Hylan": 331. "The Reminiscences of John F. O'Ryan" (1950), Oral History Collection of Columbia University, Butler Library, Columbia University, New York, NY: 110.

20. Bullock, "Hylan": 444–50. W. O. Saunders, "Red Mike or Honest

John?" *Colliers* 75 (May 9, 1925): 9. Grover A. Whalen, *Mr. New York: The Autobiography of Grover A. Whalen* (New York: G. P. Putnam's Sons, 1955): 48–59.

21. Three of the first six occasions on which Hylan's name appeared in the pages of the *New York Times* involved his criticisms of the BRT. This statement is based on an examination of the *New York Times Index,* under the heading "Hylan," from 1900 to 1916. I have not included the newspaper's coverage of Hylan's 1914 appointment to the King's County Court or his subsequent election to that court. Hylan appeared to show more animosity toward the subway companies than toward banks, realtors, railroads, gas and electric utilities, and other symbols of corporate power.

22. Mary D. Blankenhorn, "The United Brotherhood of Strap-Hangers and Coal-Diggers," *Survey* 49 (December 1, 1922): 293.

23. Irving Bernstein, *The Lean Years: A History of the American Worker, 1920–1933* (Boston: Houghton Mifflin, 1960): 63–69. Donna Gabbaccia, "Little Italy Tenements: Home, Neighborhood, Investments," a paper delivered at the Social Science Research Council Conference on the Landscape of Modernity, December 1–2, 1990, New York City, NY: 1–2, 15–21. Nancy L. Green, "Sweatshop Migrations: The Garment Industry Between Home and Shop," a paper delivered at the Social Science Research Council Conference on the Landscape of Modernity, December 1–2, 1990, New York, NY: 14–21. Mary Elizabeth Brown to Clifton Hood, March 5, 1992; in author's possession. The New York Subways Advertising Company made surveys of subway passengers starting in the 1930s, but I have been unable to locate any of them. For an insightful analysis of the popular basis of political leadership, see Alan Brinkley, *Voices of Protest: Huey Long, Father Coughlin and the Great Depression* (New York: Vintage Books, 1983): ix–xiii.

24. *New York Evening Post,* November 2, 1918.

25. *Brooklyn Daily Eagle,* November 2, 1918.

26. *New York American,* November 2, 1918.

27. Ibid.

28. *New York Evening Post,* November 2, 1918. *New York World,* November 2, 1918. *Brooklyn Daily Eagle,* November 2 and 3, 1918. *Brooklyn Citizen,* November 2, 1918. *Brooklyn Standard-Union,* November 2 and 3, 1918. *New York American,* November 2 and 3, 1918. Cunningham and DeHart, *A History of the New York City Subway System,* Part 2, *Rapid Transit in Brooklyn:* 40, 50. John Taurnac, "Out of Service!!" *Seaport* 25 (Summer 1991): 34–39. Brian Cudahy, *Under the Sidewalks of New York: The Greatest Subway System in the World,* rev. ed. (Lexington, MA: Stephen Greene Press, 1988): 71–79. Historians disagree about the number of people who died in the wreck, and the actual number may never be known. I have accepted Brian J. Cudahy's carefully verified total as the most accurate count. Brian J. Cudahy, "The Malbone Street Wreck: A Moment in the History of Urban Transportation," a paper delivered May 5, 1991, at the Subway History Conference, New York Transit Museum, Brooklyn, NY.

29. *Brooklyn Standard-Union,* November 2, 1918. No BRT employee was ever punished for the accident. Edward Luciano was tried and acquitted on a

manslaughter charge. Five other BRT officials were acquitted, too. The Brooklyn Rapid Transit Company itself paid about $2.2 million to survivors and heirs. Cudahy, *Under the Sidewalks of New York:* 79.

30. Cudahy, *Under the Sidewalks of New York:* 79.

31. *Brooklyn Daily Eagle,* November 2, 1918. *Brooklyn Standard-Union,* November 2 and 3, 1918.

32. These reorganization plans involved increases in the rate of fare paid by the passengers; reductions in the taxes, licensing fees, and interest and rental payments due the government; and the devaluation of stocks and bonds held by investors. *Commercial & Financial Chronicle* 113 (October 1, 1921): 1431–34. *Commercial & Financial Chronicle* 113 (October 15, 1921): 1646–47. "Radical Readjustment Proposed in New York City," *Electric Railway Journal* 58 (October 1, 1921): 557–60. *New York Times,* January 4, July 20 and 25, 1919; April 9, 1920. New York's transit regulatory commissions were reorganized several times following World War I, with the main conflict involving the issue of Republican/upstate influence versus Democratic/home rule control. The New York State Transit Construction Commission (TCC) assumed the power to promote new rapid transit lines from the New York State Public Service Commission (PSC) for the First District in 1919, during Governor Alfred E. Smith's administration; in 1921, the New York State Transit Commission (TC) replaced both the TCC and the PSC and acquired from them the authority to build new rapid transit lines and to regulate existing transit lines. In 1924 the New York City Board of Transportation, in turn, took over the power to develop new rapid transit lines from the TC.

33. *New York Times,* May 15, 1922. "Statement by John F. Hylan on Jenks Bills Nos. 1961, 1962, and 1963," n.d., State Legislature—Bills—Transit folder, box 355, Mayor John F. Hylan Papers, Municipal Archives, New York City Department of Records and Information Services, New York, NY.

34. John F. Hylan, "Traction and Finance," *Forum* 65 (March 1921): 258. See also Derrick, "The New York City Transit Crisis of 1918–1925": 32–59.

35. New York State Transit Commission, *Report of the Special Counsel to the Transit Commission, Metropolitan Division, Department of Public Service of the State of New York, on the Definitive Plan and Unification Agreements, Proposed for the Acquisition and Control of Rapid Transit Railroads and Related Power Properties in the City of New York, dated June 22, 1936, together with Alternative Recommendations and Memorandum of Law* (New York: n.p., 1937): 36–37.

36. *New York Times,* August 30, 1922.

37. Robert D. Cuff, *The War Industries Board: Business-Government Relations During World War I* (Baltimore: Johns Hopkins University Press, 1973): 8–12, 264–76. William E. Leuchtenburg, "The New Deal and the Analogue of War," in *Change and Continuity in the Twentieth-Century,* eds. John Braeman, Robert H. Bremner, and Everett Walters (New York: Harper & Row, 1966): 81–143. Ellis W. Hawley, "Herbert Hoover, the Commerce Secretariat, and the Vision of an 'Associative State,' " *Journal of American History* 61 (June 1974): 116–40. Michael N. Danielson and Jameson W. Doig, *New York: The Politics of Urban Regional Development* (Berkeley, CA: University of California Press,

1982): 186–94. Erwin Wilkie Bard, *The Port of New York Authority* (New York: Columbia University Press, 1939): 3, 17, 49. David C. Hammack, "Private Organizations, Public Purposes: Nonprofits and the Archives," *Journal of American History* (June 1989): 181–91. David C. Hammack, "Political Sources of New York's Modern Cityscape, 1900–1933," a paper delivered at the Social Science Research Council Conference on the Landscape of Modernity, December 1–2, 1990, New York, NY: 6–14.

38. Keith D. Revell, "Regulating the Landscape in New York City: Real Estate Values, City Planning, and the 1916 Zoning Ordinance," a paper delivered at the Social Science Research Council Conference on the Landscape of Modernity, December 1–2, 1990, New York, NY: 26. I am grateful to Jameson W. Doig for informing me about Hylan's opposition to the Port Authority.

39. John F. Hylan, *Address to the Ninth Annual Conference of Mayors and Other City Officials of the State of New York, Newburgh, N.Y., June 13, 1918* (New York: n.p., 1918): 2–3. Hylan, "Traction and Finance," *Forum:* 257–66. *Mayor Hylan's Plan for Real Rapid Transit in New York City* (New York: n.p., 1922): 6–20. In 1940 the City of New York did succeed in unifying the IRT and BMT. However, the price would have been much higher in the early 1920s partly because of the impact of the depression and partly because of the depreciation of the subways. As discussed in chapters 9 and 10, unification did not prove to be a panacea.

40. Quoted in "The Traction Crisis in New York," *Review of Reviews* 61 (March 1920): 330.

41. One exception to the nickel fare was the Fifth Avenue Coach Company, which charged ten cents for a ride on its buses; this company was universally understood to offer premium service. See also Peter Derrick, "The N.Y.C. Mess: Legacy of the 5¢ Fare," *Mass Transit* 8 (July 1981): 12–13, 26.

42. Hylan, "Traction and Finance": 258.

43. John F. Hylan to voters, August 27, 1925, John F. Hylan Correspondence file, box A-20, Citizens Union Collection, Rare Books and Manuscript Library, Butler Library, Columbia University, New York, NY.

44. Quoted in "Hylan's Hold on New York," *Literary Digest* 71 (November 19, 1921): 12.

45. "Feeding the Mob," *Saturday Evening Post* 200 (August 20, 1927): 26.

46. *National Cyclopaedia of American Biography,* v. E (New York: James T. White and Company, 1938): 77–78. *New York Times,* March 13, 1942.

47. Daniel L. Turner, "Rapid Transit Development," in New York City Board of Estimate and Apportionment, Committee on City Plan, *Development and Present Status of City Planning in New York City* (New York: n.p., 1914): 50–51.

48. Ibid.: 50–53. See also Daniel L. Turner, "Increase in Subway Congestion," *Real Estate Record and Builders Guide* (July 8, 1916): 41–42, and Daniel L. Turner, "How Can New York's Transit Problem Be Solved?" *Electric Railway Journal* 59 (February 18, 1922): 281–83.

49. New York State Construction Commissioner, *Proceedings of the Transit Construction Commissioner, June 1 to December 31, 1919,* v. 2 (New York: n.p., 1920): 916–34. The transit lines that Turner wanted to build in midtown were moving platforms. A moving platform consisted of three conveyor belts

that ran at different speeds, starting at three miles per hour on the inside belt, where passengers climbed aboard, increasing to six miles per hour in the middle, and then to nine miles per hour on the inside.

50. Ibid.: 916–35.

51. Ibid.: 920.

52. Ibid.: 919.

53. David Laird, "Versions of Eden: The Automobile and the American Novel," *Michigan Quarterly Review* 19–20 (Fall 1980–Winter 1981): 639–51. Gerald D. Silk, "The Image of the Automobile in American Art," *Michigan Quarterly Review* 19–20 (Fall 1980–Winter 1981): 601–16. General Motors, *The General Motors Exhibit Building* (New York: n.p., 1939): unpaginated. General Motors, *Futurama* (New York: n.p., 1939): unpaginated. See also "Catalogue of Exhibitions": 101, and Joseph P. Cusker, "The World of Tomorrow: Science, Culture and Community at the New York World's Fair," both in *Dawn of a New Day: The New York World's Fair 1939/1940,* ed. Helen A. Harrison (New York: Queens Museum, 1980): 10, 14; and Blaine A. Brownell, "A Symbol of Modernity: Attitudes Toward the Automobile in Southern Cities in the 1920s," *American Quarterly* 24 (March 1972): 20–44.

54. Foster, *From Streetcar to Superhighway:* 38–45. Bottles, *Los Angeles and the Automobile:* 211–15.

55. Regional Survey of New York and Its Environs, *Regional Survey of New York and Its Environs,* v. 3, *Highway Traffic* (New York: Regional Plan of New York and Its Environs, 1927): 127–34. The Regional Plan also recommended railroad improvements. See Regional Survey of New York and Its Environs, *Regional Survey of New York and Its Environs,* v. 4, *Transit and Transportation* (New York: Regional Plan of New York and Its Environs, 1928): 18. Robert Fishman, "The Regional Plan of New York and Its Environs (1929–31) and the Transportation of the Industrial Metropolis," a paper delivered at the Social Science Research Council Conference on the Landscape of Modernity, December 1–2, 1990, New York, NY: 5–6.

56. *New York World,* May 11, 1922. Regional Survey of New York and Its Environs, *Regional Survey of New York and Its Environs,* v. 4, *Transit and Transportation:* 16–21, 129–30, 178–96. Fishman, "Regional Plan": 5–6.

57. Robert A. Caro, *The Power Broker: Robert Moses and the Fall of New York* (New York: Vintage Books, 1974): 11, 897–98, 909, 912, 934.

58. Regional Survey of New York and Its Environs, *Regional Survey of New York and Its Environs,* v. 4, *Transit and Transportation:* 17; Fishman, "Regional Plan": 13.

59. Fishman, "Regional Plan": 19–24.

60. Jameson W. Doig, "Joining New York City to the Greater Metropolis: the Port Authority as Visionary, Target of Opportunity, and Opportunist," a paper delivered at the Social Science Research Council Conference on New York's Built Environment, December 1–2, 1991, New York, NY: 19–24. Caro, *The Power Broker:* 13–18, 625–40.

61. Kenneth T. Jackson, *The Crabgrass Frontier: The Suburbanization of the United States* (New York: Oxford University Press, 1985): 3–11, 169–86, 290–96.

62. *Mayor Hylan's Plan for Real Rapid Transit* (New York: n.p., 1922): 6.

63. Ibid.: 7.

64. Ibid.: 3.

65. New York State Transit Commission, *New Subways: Proposed Additions to Rapid Transit System to Cost $218,000,000* (New York: n.p., 1922): 1–16. *New York Sun,* May 11, 1922.

66. *New York Times,* May 18, 1922. See also Derrick, "The New York City Transit Crisis of 1918–1925": 50–60.

67. Alfred E. Smith, *Up to Now: An Autobiography* (New York: Viking Press, 1929): 155–56, 332–33. Richard O'Connor, *The First Hurrah: A Biography of Alfred E. Smith* (New York: G. P. Putnam's Sons, 1970): 95–97, 109, 110, 158–61. Oscar Handlin, *Al Smith and His America* (Boston: Little, Brown and Company, 1958): 70–73.

68. *New York Times,* January 12, 1936. "The Smith-Hylan Battle," *Literary Digest* 86 (September 12, 1925: 8–9. "Smith's Triumph over Hylan and Hearst," *Literary Digest* 86 (September 26, 1925): 8–9. See also Peter Windley Herman, "An Historical Examination of the City-State Controversy over the Control of the New York City Subways," *Urban Lawyer* 3 (Winter 1971): 78–98.

69. New York State, *Public Papers of Alfred E. Smith, Forty-Seventh Governor of the State of New York, 1924* (Albany: J. B. Lyon and Company, 1926): 46–48. *New York Times,* February 7, 1924. Warren Moscow, *The Last of the Big-Time Bosses: The Life and Times of Carmine De Sapio and the Rise and Fall of Tammany Hall* (New York: Stein and Day, 1971): 22–23. Martin Landau, "The New York Public Service Commission, 1907–1930: A Study of Regulation in Its Political Environment" (Ph.D. dissertation, New York University, 1952): 84.

70. *New York Times,* February 1 and November 1, 1922; February 20, April 4, 5, 7, 9, 11, 12, and May 3, 1924; August 15, 1952. New York State, *Laws of the State of New York,* v. 2 (Albany: J. B. Lyon and Company, 1924): 1038–46. Joel Fischer, "Urban Transportation: Home Rule and the Independent Subway System in New York City, 1917–1925" (Ph.D. dissertation, St. John's University, 1978): 244–56. The self-sustaining fare was never imposed. During the 1930s, the state legislature kept extending the "temporary" adjustment period when the nickel fare was in effect, and the turnstile cost was never raised.

71. Daniel L. Turner, "Memorandum for Chairman McAneny Re. Transit Commission New Subway Lines," January 11, 1927, Transit 1927 folder, box 35, George McAneny Papers, Seely G. Mudd Manuscript Library, Princeton University, Princeton, NJ; NYCBT, *Communication from the Board of Transportation to the Board of Estimate Transmitting Proposed Form of Contract and Lease for Operation and Maintenance of Independent City Owned Rapid Transit Railroad* (New York: n.p., 1932): 12–13.

72. NYCBT, *Communication from the Board of Transportation:* 12–13; Peter Derrick, "Catalyst for Development: Rapid Transit in New York," *New York Affairs* 9 (Fall 1986): 29–76.

73. Barrett, *The Automobile and Urban Transit:* 3–8. *New York Times,* December 10, 1924; April 9, 1936. NYSTC, *New Subways:* 1–16. For the replacement of elevateds by subways, see George W. Seaton, *Cue's Guide to New*

York City (New York: Prentice-Hall, 1940): 17–18; *Commercial and Financial Chronicle* 116 (May 12, 1923): 2130; Cunningham and DeHart, *A History of the New York City Subway System,* Part 1, *The Manhattan Els and the IRT:* 68; Cunningham and DeHart, *A History of the New York City Subway System,* Part 2, *Rapid Transit in Brooklyn:* 69–71; Walter D. Binger to Stanley M. Isaacs, December 30, 1940, Transit Elevated Structures—Removal of 2nd and 3rd Avenue Els folder, box 2590, FHL. For the related issue of the replacement of buses by trolleys, see Clifton Hood, "Underground Politics: A History of Mass Transit in New York City Since 1904" (Ph.D. dissertation, Columbia University, 1986): 298–302; David J. St. Clair, "The Motorization and Decline of Urban Public Transit, 1935–1960," *Journal of Economic History* 41 (September 1981): 579–600; and John B. Rae, "The Evolution of the Motor Bus as a Transport Mode," *High Speed Ground Transportation Journal* 5 (Summer 1971): 221–35. For Hylan's support of buses over trolleys, see *New York American,* March 22, 1922.

74. Robert A. M. Stern, Gregory Gilmartin, and Thomas Mellins, *New York 1930: Architecture and Urbanism Between the Two World Wars* (New York: Rizzoli, 1987): 70, 307.

75. *New York Times,* December 11 and 19, 1924.

76. Travis H. Whitney, "Transit in New York City," a speech delivered on January 14, 1925 at Congregation Beth Elohin, New York, NY. See also "Possible City Operation for New York, but with Service at Cost," *Electric Railway Journal* 63 (April 19, 1924): 203, and George F. Swain, *Report on Rapid Transit Situation in New York City* (New York: M. B. Brown, 1927): 9–10.

77. *New York Times,* December 11, 1926.

78. Merchants' Association, Committee on City Transit, *Subway Consolidation Necessary for Adequate Service and Economy* (New York: n.p., 1927): 5.

79. *New York Times,* March 10, 1925.

80. *New York Times,* December 10 and 11, 1925.

81. NYCBT, *Proceedings,* v. 1 (New York: n.p., 1925): 539–41, 631–32. NYCBT, *Proceedings,* v. 2 (New York: n.p., 1926): 523–59. NYCBT, *Proceedings,* v. 3 (New York: n.p., 1926): 899–901. New York State, *Laws of the State of New York, 1924:* 104–42. Fischer, "Urban Transportation": 244–56. NYCBT, *Communication from the Board of Transportation to the Board of Estimate:* 12–13, 30–37.

82. *New York Times,* March 15, 1925.

83. NYCBT, *Communication from the Board of Transportation to the Board of Estimate:* 12. For the construction of the IND, see Frederick A. Kramer, *Building the Independent Subway: The Technology and Intense Struggle of New York's Most Gigantic Venture* (New York: Quadrant Press, 1990): 16–80.

Chapter 9. The People's Subway, the Nickel Fare, and Unification

1. For the automobile, see David Laird, "Versions of Eden: The Automobile and the American Novel," *Michigan Quarterly Review* 19–20 (Fall 1980–

Winter 1981): 639–51; Gerald D. Silk, "The Image of the Automobile in American Art," *Michigan Quarterly Review* 19–20 (Fall 1980–Winter 1981): 601–16; and Blaine A. Brownell, "A Symbol of Modernity: Attitudes Toward the Automobile in Southern Cities in the 1920s," *American Quarterly* 24 (March 1972): 20–44. For the airplane, see Joseph J. Corn, *The Winged Gospel: America's Romance with Aviation, 1900–1950* (New York: Oxford University Press, 1983): 3–27, and Roger E. Bilstein, *Flight in America, 1900–1930: From the Wrights to the Astronauts* (Baltimore: Johns Hopkins University Press, 1984): 23–34, 60–64, 76–77.

2. Christopher Morley, "Thoughts in the Subway," in *Christopher Morley's New York* (New York: Fordham University Press, 1988): 116.

3. Josephine Gear, *Straphangers* (New York: Whitney Museum of American Art, 1989): unpaginated. Mary Ryan Gallery, *The Artist and the El* (New York, 1982): unpaginated. Rodger C. Birt, "An Analysis of the New York Transit Museum's Photographic Documentation of Project One," a paper presented May 10, 1991, at the Subway History Conference, New York Transit Museum, Brooklyn, NY.

4. *New York Times,* September 13, 1926; August 16, 1934. Robert Sklar, *Movie-Made America: A Social History of American Movies* (New York: Random House, 1975): 161–94. Warner Brothers, *Dames,* 1934. William Saroyan, *Subway Circus* (New York: Samuel French, 1940): 5.

5. Interview with Gunther Eichholz, March 31, 1991, New York, NY.

6. Stanley M. Isaacs, "Statement for WMCA on Five Cent Fare," WMCA Business Forum Broadcast No. 37, A New Era for the Straphanger, November 24, 1943, box 10, SMI. In 1942, City Councilman Peter V. Cacchione estimated that a ten-cent fare would impose on the average manufacturing worker the equivalent of a 1.5 percent income tax. Joshua Freeman, *In Transit: The Transport Workers Union in New York City* (New York: Oxford University Press, 1989): 288.

7. Herman Wouk, *The City Boy* (New York: Simon & Schuster, 1948): 55–65. See also E. L. Doctorow, *World's Fair* (New York: Fawcett Crest, 1985): 148, 156–59.

8. *New York Times,* March 25, 1973. See also Alfred Kazin, *A Walker in the City* (New York: Harcourt, Brace & World, 1951): 5–47.

9. *New York Times,* October 18, 1985.

10. "New York in the Eighties—a Symposium," *New Criterion* (Summer 1986): 60.

11. Vermont Royster, *My Own, My Country's Time: A Journalist's Journey* (Chapel Hill, NC: Algonquin Books, 1983): 45.

12. *Wall Street Journal,* January 30, 1986.

13. Odette Keun, *I Think Aloud in America* (London: Longmans, Green and Company, 1938): 49–50. See also "An Epic of the Subway," *Living Age* 336 (May 1929): 217; Haruko Ichikawa, *Japanese Lady in New York* (Tokyo: Kenkyusha Company, 1938): 50; and A. G. MacDonnel, *A Visit to America* (New York: Macmillan, 1935): 21. For other accounts that emphasize the dehumanization of subway riders, see Seymour Gross, *Subway Face: An Expressionistic Play* (Ithaca, NY: n.p., 1935), and Elmer Rice, *The Subway: A Play in Nine Scenes* (New York: n.p., 1929).

14. MacDonnel, *A Visit to America:* 48–50. *New York Times,* May 4, 1929; November 26, 1927; January 10, 1932; April 6 and June 8, 1933; January 16, 1934; February 21, 1935; March 22 and 27, 1937. "End of Mr. Badger," *Atlantic Monthly* 161 (June 1938): 844–47. Laurence Bell, "The Most Awful Ride in the World," *American Mercury* 45 (October 1938): 142–49. A. L. Merritt, "Human Nature as Seen in the New York Subway," *American* 92 (September 1921): 48–49, 112–13. For a later account that romanticizes the supposedly bohemian life of the homeless in the subway, see Edmund G. Love, *Subways Are for Sleeping* (New York: Harcourt, Brace and Company, 1956). See also Rick Beard, ed., *On Being Homeless: Historical Perspectives* (New York: Museum of the City of New York, 1987).

15. *New York Times,* May 15 and 17, 1936. NYCBT, "Abridged Report of Occurrences on the Independent Subway System During the Week Ended Saturday, October 17, 1936," October 19, 1936, Board of Transportation folder, box 675, FHL. NYCBT, "Abridged Report of Occurrences on the Independent Subway System for the Week Ended March 6, 1937," March 8, 1937, Board of Transportation folder, box 681, FHL. NYCBT, "Abridged Report of Occurrences on the Independent Subway System During the Week Ended Saturday, July 3, 1937," July 6, 1937, Board of Transportation folder, box 682, FHL. NYCBT, "Abridged Report of Occurrences on the Independent Subway System During the Week Ended Saturday, December 24, 1937," December 27, 1937, Board of Transportation folder, box 680. NYCBT, "Abridged Report of Occurrences on the Independent Subway System During the Week Ended Saturday, January 20, 1940," January 22, 1940, Board of Transportation folder, box 698, FHL. "Arrests Made by I.R.T. Special Officers for the Year Ended December 31, 1938," Transit Commission folder, box 694, FHL. William G. Fullen to Henry H. Curran, February 25, 1939, Transit Commission folder, box 694, FHL.

16. *New York American,* February 3, 1921. George McAneny, undated press release, box 33, George McAneny Papers, Seely G. Mudd Manuscript Library, Princeton University, Princeton, NJ. George McAneny to Judge Abel E. Blackmar, May 29, 1925, Miscellaneous Transit folder, box 35, George McAneny Papers. George McAneny, Notes on Transit Unification, September 11, 1934, Transit 1934 folder, box 44, George McAneny Papers. August Belmont to Messrs. de Rothschild Frères, July 23, 1921, CL 6, BFP.

17. *Gilchrist v. Interborough Rapid Transit Company,* 279 U.S. 159 (1929). "The Nickel Wins in the Supreme Court," *Literary Digest* 101 (April 20, 1929): 13. *Commercial and Financial Chronicle* 126 (May 5, 1928): 2791. Cynthia M. Latta, "The Return on the Investment on the Interborough Rapid Transit Company" (Ph.D. dissertation, Columbia University, 1974): 271–78.

18. Thomas F. Woodlock, "New York's Struggle to Maintain Five-Cent Fare," *Barron's* 12 (March 2, 1932): 12. Latta, "The Return on the Investment": 230–35. Peter Derrick, "The N.Y.C. Mess: Legacy of the 5¢ Fare," *Mass Transit* 8 (July 1921): 12–13, 26. For proposals to reform the fare structure, see William Vickrey, "The Revision of the Rapid Transit Fare Structure of the City of New York," Finance Project, Mayor's Committee on Management Survey of the City of New York, Technical Monograph No. 3 (New York: n.p., 1952), and William Vickrey, *Fare Structure Reform for New York City* (New York:

n.p., 1981). For critiques of the argument for low fares, see Dick Netzer, "The Case Against Low Subway Fares," *New York Affairs* 1 (Winter 1974): 14–25, and David Stewart, "Rolling Nowhere," *Inquiry* 7 (July 1984): 18–23.

19. Interborough Rapid Transit Company, *Annual Report,* 1920–1930 (New York: n.p., 1920–30). Brooklyn-Manhattan Transit Corporation, *Annual Report,* 1923–1930 (New York: n.p., 1920–30).

20. *New York Times,* June 28, 1922. August Belmont to S. Stephany, July 23, 1921, CL 6, BFP. See also Joshua B. Freeman, *In Transit: The Transport Workers Union in New York City, 1933–1966* (New York: Oxford University Press, 1989): 3–35, and Latta, "The Return on the Investment": 259–66.

21. August Belmont to S. Stephany, November 29, 1920, CL 6, BFP.

22. *New York Times,* December 8, 1924; June 14, 1928. August Belmont to S. Stephany, November 29, 1920, CL 6, BFP. August Belmont to Messrs. de Rothschilds Frères, July 22, 1921, CL 6, BFP. "Multiple Unit Door Control on the Interborough Subway Trains," *Electric Railway Journal* 66 (September 19, 1925): 429–33. Latta, "The Return on the Investment": 265. An unintended impact of this labor-saving machinery may have been to facilitate the rise of crime in the 1960s and afterward. Despite New York City Transit Authority's efforts to combat crime through increases in the size of its police force, the number of employees who work in the stations and trains is less in the 1990s than it was in the 1920s. The presence of several ticket choppers in every station and of five guards per ten-car train ensured a regular uniformed presence and presumably had a deterrent effect. If earlier manning levels had been continued, the subways may have been less susceptible to the crime and disorderliness that arose after the 1950s. See Clifton Hood, "Underground Politics: A History of Mass Transit in New York City Since 1904" (Ph.D. dissertation, Columbia University, 1986): 445–52.

23. *New York Times,* January 31, 1927; February 26, 1927. August Belmont to Messrs. de Rothschild Frères, July 22, 1921, CL 6, BFP. E. Alfred Seibel, "The Golden Rod I.R.T. 'L' Cars," *ERA Bulletin* 31 (March 1988): 7.

24. Henry F. Pringle, "Wheels in His Head," *The New Yorker* 3 (December 31, 1927): 19.

25. E. Stewart Fay, *Londoner's New York* (London: Methuen and Company, 1936): 143. See also Anthony Armstrong Willis, *Britisher on Broadway* (London: Methuen and Company, 1932): 43–49.

26. NYSTC, *Nineteenth Annual Report, 1939* (Albany: Williams Press, 1942): 233–35. The IRT's elevated division was hit particularly hard, losing 40 percent of its ridership over that ten-year period, compared to only 11 percent for the company's subway division.

27. Latta, "The Return on the Investment": 41–48, 245–51, 259–63, 274–79. Under an agreement made in May 1922 under the supervision of U.S. Circuit Court Judge Julius M. Mayer, the Manhattan rental was temporally reduced from the original fixed guaranteed rental of 7 percent to 3 percent in 1922 and then to 4 percent in 1923, before going to 5 percent. *New York Times,* May 2, 1922. NYSTC, "Memorandum on Reorganization of Interborough Rapid Transit Company," December 9, 1924, Transit folder, box 33, McAneny Papers. Interborough Rapid Transit Company, *Annual Report,* 1927–1939 (New York:

n.p., 1927–39). Brooklyn-Manhattan Transit Corporation, *Annual Report,* 1929–1939 (New York: n.p., 1929–39).

28. Nathan L. Amster to Fiorello H. LaGuardia, February 8, 1938, Manhattan Railway Municipal Taxes folder, box 2590, FHL. Nathan L. Amster to Fiorello H. LaGuardia, June 29, 1938, Transit—Sixth Avenue "L" Demolition folder, box 2590, FHL. Fiorello H. LaGuardia, press release, March 29, 1938, Manhattan Railway Municipal Taxes folder, box 2590, FHL. Fiorello H. LaGuardia to William G. Fullen, May 23, 1934, Transit Matters folder, box 2590, FHL.

29. Thomas Kessner, *Fiorello H. LaGuardia and the Making of Modern New York* (New York: McGraw-Hill, 1989): 258.

30. Ibid.: 3–33. Arthur Mann, *LaGuardia: A Fighter Against His Times, 1882–1933* (Chicago: University of Chicago Press, 1959): 22–24, 26–28, 33–39, 49–54.

31. Arthur Mann, *LaGuardia Comes to Power: 1933* (Chicago: University of Chicago Press, 1965): 121, 124. Kessner, *Fiorello H. LaGuardia and the Making of Modern New York:* 192–95, 222–253. Herbert Mitgang, *The Man Who Rode the Tiger: The Life and Times of Judge Samuel Seabury* (Philadelphia: J. B. Lippincott, 1963): 162–165, 313–25.

32. William Manners, *Patience and Fortitude: Fiorello LaGuardia* (New York: Harcourt Brace Jovanovich, 1976): 208.

33. Arthur Mann, "Fiorello Henry LaGuardia," *Dictionary of American Biography,* v. 24, ed. John A. Garraty and Edward T. Jones (New York: Charles Scribner's Sons, 1974): 464–67. Kessner, *Fiorello H. LaGuardia and the Making of Modern New York:* xi–xvi, 255–409. Mark I. Gelfand, *A Nation of Cities: The Federal Government and Urban America, 1933–1965* (New York: Oxford University Press, 1975): 43–48, 380–89.

34. Oral history interview with Goodhue Livingston, 10, November 12, 1982, LaGuardia and Wagner Archives, LaGuardia Community College/CUNY, Long Island City, NY.

35. *New York American,* February 2 and 3, 1921. *New York Evening Post,* June 1, 1921. *Speaker's Handbook: Eight Years of the LaGuardia Administration, 1934* (New York: n.p., 1941): 20–24, 32–34. *The Fusion Handbook* (New York: n.p., 1933): 118–20. Fiorello H. LaGuardia, "Suggested Outline: City of Tomorrow, 1947," folder 17, box 24B1, LaGuardia Papers, LaGuardia and Wagner Archives, LaGuardia Community College/CUNY, Long Island City, NY. Kessner, *Fiorello H. LaGuardia and the Making of Modern New York:* 69–75, 432–34, 460. Fiorello H. LaGuardia, "The Demolition of the Sixth Avenue Elevated," a transcript of a broadcast over radio station WJZ, December 20, 1935, Transit Unification—"L" Demolition, Sixth Avenue "L" folder, box 773, FHL. Fiorello H. LaGuardia to Herbert H. Lehman, February 15, 1941, New York State—Correspondence with Governor Lehman, 1941–42 folder, box 2566, FHL. Fiorello H. LaGuardia to Eleanor E. Sherman, August 31, 1921, "L" Removal—Third Avenue folder, box 773, FHL. Mayor's Traffic Committee, Subcommittee on Buses and Taxicabs, *Findings and Recommendations,* c. 1937, Traffic Committee, Mayor's folder, box 2589, FHL; Fiorello H. LaGuardia, "LaGuardia: Plot to Gag Straphangers," *PM,* December 1, 1946. Fior-

ello H. LaGuardia, "LaGuardia: Don't Fall for a 10-Cent Fare," *PM,* February 9, 1947. Fiorello H. LaGuardia, transcript of broadcast over radio station WJZ, November 24, 1946, folder 14, box 26D2, LaGuardia Papers, LaGuardia and Wagner Archives, LaGuardia Community College/CUNY, Long Island City, NY.

36. Oral history interview with Goodhue Livingston: 10, November 12, 1982.

37. Robert Moses to Thomas E. Dewey, August 30, 1968, series 8, box 25, folder 10, Thomas E. Dewey Papers, Department of Rare Books and Special Collections, Rush Rhees Library, University of Rochester, Rochester, NY.

38. *New York Times,* May 27, 1919; August 15, 1952.

39. Kessner, *Fiorello H. LaGuardia and the Making of Modern New York:* 48, 69–75, 432–35, 460. Mann, *LaGuardia: A Fighter Against His Times:* 81, 227, 290. *Speaker's Handbook: Eight Years of the LaGuardia Administration, 1934:* 20–24, 32–34. *The Fusion Handbook:* 118–20. LaGuardia, "Suggested Outline: City of Tomorrow, 1947." LaGuardia thought that the els marred the urban landscape with their ugly columns and risers, retarded real estate development, and interfered with pedestrian and vehicular traffic. LaGuardia, "The Demolition of the Sixth Avenue Elevated." LaGuardia to Lehman, February 15, 1941. LaGuardia to Sherman, August 31, 1921. LaGuardia also disparaged street railways as obsolete throwbacks that had no place in the modern city and once asserted that "trolley cars were as dead as sailing ships." Fiorello H. LaGuardia, address identified as "1935 speech," Buses 1934 folder, box 726, FHL.

40. Robert A. Caro, *The Power Broker: Robert Moses and the Fall of New York* (New York: Vintage Books, 1975): 358–60, 445–53, 462–63, 607–14, 932–33. Kessner, *Fiorello H. LaGuardia and the Making of Modern New York:* 300–9.

41. For accounts of the unification of London's private transit companies under public control through the passage of the London Passenger Transport bill of 1933, see E. Edon Barry, *Nationalism in British Politics: The Historical Background* (Stanford, CA: Stanford University Press, 1965): 284–303, and T. C. Barker and Michael Robbins, *A History of London Transport,* v. 2, *The Twentieth Century to 1970* (London: George Allen and Unwin, 1976): 270–311, 337–49.

42. Kessner, *Fiorello H. LaGuardia and the Making of Modern New York:* 262.

43. Ibid.: 215–22, 262–69.

44. City Club of New York, *The Untermyer Plan of Subway Unification* (New York: n.p., 1931): 1–6.

45. NYCBT, *Communication from Board of Transportation to the Board of Estimate and Apportionment of the City of New York, Transmitting Proposed Form of Contract and Lease for Operation and Maintenance of Independent Rapid Transit Railroad* (New York: n.p., 1932): 3–37. NYCBT, *Report of the Independent City-Owned Rapid Transit for the Fiscal Year Ended June 30, 1939* (New York: n.p., 1939): 4–6. Arthur J. Waterman, Jr., "The Integration of Rapid Transit Facilities of the City of New York" (Ph.D. dissertation, New

York University, 1940): 250–71. NYCBT, *Proceedings, January 1, 1932 to June 30, 1932,* v. 18 (New York: n.p., 1932): 452, 497–98, 547–48, 833–34, 1164, 1478–88. Although the 1924 law creating the Board of Transportation had stipulated that a municipally run subway must have a self-sustaining fare after an initial three-year trial period, the Democratic legislatures of the 1930s repeatedly extended the length of this trial period, effectively nullifying this condition. Waterman, "The Integration of Rapid Transit Facilities": 270–71.

46. NYSTC, *Report of Special Counsel on the Definitive Plan and Unification Agreement, Proposed for the Acquisition and Unification under Public Ownership and Control of Rapid Transit Railroads and Related Power Properties in the City of New York, Dated June 22, 1936, Together with Alternative Recommendations and Memorandum of Law* (New York: n.p., 1937): 9. Waterman, "The Integration of Rapid Transit Facilities": 267.

47. NYCBT, *Report of the Independent City-Owned Rapid Transit Railroad 1932/34–1935/36* (New York: n.p., 1936): 11. NYCBT, *Report of the Independent City-Owned Rapid Transit Railroad for the Fiscal Year Ended June 30, 1937* (New York: n.p., 1937): 6–9. NYCBT, *Report of the Independent City-Owned Rapid Transit Railroad for the Fiscal Year Ended June 30, 1938* (New York: n.p., 1938): 6–11. NYCBT, *Report of the Independent City-Owned Rapid Transit Railroad for the Fiscal Year Ended June 30, 1939* (New York: n.p., 1939): 8–9. NYCBT, *Report of the Independent City-Owned Rapid Transit Railroad for the Fiscal Year Ended June 30, 1940* (New York: n.p., 1940): 8–9. NYCBT, *Annual Report of the Secretary, 1932* (New York: n.p., 1933): 8–9. "With Doubt and Dismay," *Transit Journal* 76 (July 1932): 301–2.

48. Adolf A. Berle to Fiorello H. LaGuardia, Samuel Seabury, and Paul Windels, "Memorandum re. Transit Unification—Securities to Be Exchanged," February 23, 1934, Transit—Seabury-Berle Plan—Hearing and Disposition folder, box 2590, FHL. *New York Times,* January 2 and 7, 1934.

49. "The Unified Transit Plan," *Bradstreet's* 59 (January 3, 1931): 5. "Untermyer Transit Changes Likely," *Barron's* 11 (April 13, 1931): 13. Samuel F. Howard, Jr., "Samuel Untermyer," *Dictionary of American Biography,* v. 22, ed. Robert Livingston Schuyler and Edward T. James (New York: Charles Scribner's Sons, 1958): 174–76. "Transit Board's Unification Plan Meets Opposition," *Barron's* 11 (December 28, 1931): 7. Samuel Untermyer, *Report and Recommendation of Mr. Samuel Untermyer on Transit, and His Letter of Resignation as Special Counsel* (New York: n.p., 1931): 1–22. Waterman, "The Integration of Rapid Transit Facilities": 351–56.

50. This grant and loan was increased to $25,349,000 in November 1935. It did not cover the Sixth Avenue line. Waterman, "The Integration of Rapid Transit Facilities": 267. Gelfand, *A Nation of Cities: The Federal Government and Urban America, 1933–1965:* 230–35.

51. Paul Barrett, *The Automobile and Urban Transit: The Formation of Public Policy in Chicago, 1900–1930* (Philadelphia: Temple University Press, 1983): xi–8. James Blaine Walker, "Transit Unification in New York," *Public Utilities Fortnightly* 19 (February 18, 1937): 222–31. Berle, "Memorandum re. Transit Unification."

52. Joshua Freeman, "From Private to Public: Mayor LaGuardia and

the Subways," Catalog essay (New York: n.p., 1990): 20–22. *New York Times,* January 7, 1934. Fiorello H. LaGuardia to William G. Fullen, May 23, 1934, Transit Matters folder, box 2590, FHL. William G. Fullen to Fiorello H. LaGuardia, May 29, 1934, Transit Matters folder, box 2590, FHL. Fiorello H. LaGuardia to William G. Fullen, July 8, 1937, Transit Matters folder, box 2590, FHL.

53. Mitgang, *The Man Who Rode the Tiger:* 6–14, 19, 29–33, 62–63, 80–93, 145, 267. Samuel Seabury, *Municipal Ownership and Operation of Public Utilities in New York City* (New York: Municipal Ownership Publishing Company, 1905): 7–8, 64–67. Richard O. Boyer, "Inquisitor," *The New Yorker* 7 (June 27, 1931): 20.

54. Jordan A. Schwarz, *Liberal: Adolph A. Berle and the Vision of an American Era* (New York: Free Press, 1987): 96.

55. Ibid.: 1–77, 91–103. *New York Times,* February 19, 1971. Adolf A. Berle and Gardiner C. Means, *The Modern Corporation and Private Property* (New York: Macmillan, 1932): 1–10. "A Brain Trust at Work," *Review of Reviews and World's Week* 88 (July 1933): 20.

56. "The Reminiscences of Newbold Morris" (1950), Oral History Collection of Columbia University, Butler Library, Columbia University, New York, NY: 45. Interview with Arthur Mann, December 12, 1987, LaGuardia and Wagner Archives, LaGuardia Community College/CUNY, Long Island City, NY: 1. Kessner, *Fiorello H. LaGuardia and the Making of Modern New York:* 281–87, 477.

57. Freeman, "From Private to Public": 20. NYSTC, *Plan and Unification Agreement for the Acquisition and Unification, under Public Ownership and Control, of Rapid Transit Railroads and Related Power Properties in the City of New York* (New York: n.p., 1936): 40–44, 64–67, 178–85. Waterman, "The Integration of Rapid Transit Facilities": 333–35.

58. New York City Board of Estimate and Apportionment, *Report of Samuel Seabury, Special Counsel, and A. A. Berle, Jr., Chamberlain, to the Board of Estimate and Apportionment of the City of New York* (New York: n.p., 1935): 1–10. Samuel Seabury and A. A. Berle, Jr., to the Board of Estimate and Apportionment, January 24, 1936, Board of Estimate and Apportionment folder, box 672, FHL. *New York Times,* June 11, 1936; July 26, 1939; March 12, 1940. NYSTC, *Plan and Unification Agreement:* 1–8, 72–100. NYSTC, *Report of Special Counsel on the Definitive Plan and Unification Agreement, Proposed for the Acquisition and Unification under Public Ownership and Control of Rapid Transit Railroads and Related Power Properties of the City of New York, Dated June 22, 1936, Together with Alternative Recommendations and Memorandum of Law* (New York: n.p., 1937): 8–9. Caro, *The Power Broker:* 13–18, 430, 534. Waterman, "The Integration of Rapid Transit Facilities": 357–61.

59. *New York Times,* April 9 and June 22 and 23, 1936. NYSTC, *Rapid Transit Unification, Opinion of William G. Fullen* (New York: n.p., 1937): 11. William G. Mulligan, Jr., James T. Ellis, and Jeanne DeLuca, "New York's Transit Unification Keeps Marching On," *Public Utilities Fortnightly* 20 (December 23, 1937): 834. A few of the TC's specific objections did have merit. For instance, Reuben Haskell noted that Seabury-Berle's estimate of future ex-

penses made no allowance for the increases in labor costs almost certain to occur as a result of the CIO's organization of transit workers. NYSTC, *Rapid Transit Unification, Opinion of Reuben L. Haskell* (New York: n.p., 1937): 15–16, 34–35.

60. NYSTC, *Report of the Special Counsel:* 3.

61. One of the TC's claims—that Seabury-Berle's recommended purchase price far exceeded the subway companies' market value in 1937—was correct. The reason for this disparity was the disastrous impact of the 1937 recession, not the LaGuardia administration perfidy. See Hood, "Underground Politics": 344–47. NYSTC, *Report of the Special Counsel:* 21. Reuben L. Haskell, "Speech at the Transit Forum of Community Councils of the City of New York," February 2, 1938, reel 97, Governor's Papers, Herbert H. Lehman Papers, Division of Rare Books and Manuscripts, Lehman Library, Columbia University. *New York Times,* June 22, 1936; May 7, 1937. Diary of Adolf A. Berle, entry for November 10, 1937; 1937 folder, container 210, Adolf A. Berle Papers, Franklin D. Roosevelt Library, Hyde Park, NY.

62. Press release, Mayor LaGuardia's reply to the Transit Commission's report, radio broadcast, May 10, 1937, Transit Matters folder, box 2590, FHL.

63. Kessner, *Fiorello H. LaGuardia and the Making of Modern New York:* 396–419.

64. George McAneny to Herbert H. Lehman, February 23, 1938, reel 97, Governor's Papers, Lehman Papers. NYSTC press release, March 1, 1938, text of letter from Samuel G. Untermyer to William G. Fullen, February 23, 1938, reel 97, Governor's Papers, Lehman Papers. William G. Fullen to Nathan Sobel, November 21, 1941, reel 97, Governor's Papers, Lehman Papers. *New York Times,* September 9, 1938; June 29 and 30, 1939; October 3, 1971.

65. *New York Times,* May 15, 1938; April 6, 1978. Schwarz, *Liberal:* 110–17. *National Cyclopedia of American Biography,* v. I (New York: James T. White and Company, 1960): 355. Milton MacKaye, "Professor in Politics," *The New Yorker* 10 (August 11, 1934): 19–22.

66. *New York Times,* June 22, 1936; May 15, 1938.

67. Waterman, "The Integration of Rapid Transit Facilities": 363–72. *New York Times,* June 27, August 30, and November 2, 1939.

68. Waterman, "The Integration of Rapid Transit Facilities": 364–66. Freeman, "From Private to Public": 21–24. *New York Times,* June 28 and July 26, 1939. NYSTC, *Plan and Agreement of Unification and Readjustment for the Acquisition and Unification under Public Ownership and Control of Rapid Transit and Surface Railroads and Related Power Properties and Omnibus Lines of the Brooklyn-Manhattan Transit System in the City of New York* (New York: n.p., 1939): 5–19, 26–30. NYSTC, *Proposed Plan and Agreement of Unification and Readjustment for the Acquisition and Unification, Under Public Ownership and Control, of Rapid Transit Railroads and Related Properties in the City of New York of the Interborough and Manhattan Transit Systems* (New York: n.p., 1939): 8–25.

69. Waterman, "The Integration of Rapid Transit Facilities": 365–71. *New York Times,* June 28, July 26, and September 1, 1939; January 8, February 6, and March 13, 1940. One remaining hurdle that was averted involved the Transport

Workers Union. As historian Joshua Freeman observed, the lengthy unification negotiations were noteworthy for providing almost no consideration of unification's impact on labor. Formed in the 1930s with the help of communist organizers, the Transport Workers Union (CIO) was strongest on the IRT and BMT. It had less success on the publicly owned IND, partly because city workers thought civil service classification provided them with the same benefits as the union and partly because the Board of Transportation as a public agency was exempt from federal and state labor laws. The prospect of unification concerned the TWU. It feared that it might mean the loss of jobs on the IRT and BMT and the spread of the NYCBT's anti-union policies to the entire rapid transit system. Shortly before unification was completed, Mayor LaGuardia announced that the city would not adopt collective bargaining, allow a union shop, permit its workers to strike, or sign a labor contract. He also announced that some IRT and BMT workers would be laid off. In response the TWU began to prepare for a strike. Under this pressure and afraid that unification might be scuttled, LaGuardia backed down, agreeing to accept the legality of the IRT and BMT contracts except where they conflicted with public employment law. Freeman, *In Transit:* 191–223.

70. *New York Times,* June 2, 1940.

71. Ibid. *New York Herald-Tribune,* June 2, 1940. "Press release regarding B.M.T. subway—Last Train Saturday Night—from Jerry Daly," May 31, 1940, Board of Transportation folder, box 698, FHL.

72. *New York Times,* August 30, 1939; February 17, May 31, June 2, and June 13, 1940. NYSTC, *Annual Report for the Year Ended December 31, 1940* (Albany: n.p., 1941): 11–28. Freeman, "From Private to Public": 2–3. Freeman, *In Transit:* 201.

Chapter 10. The Revolt Against Politics

1. NYCBT, *Transit Record* 21 (September 1941): 1. NYCBT, *Transit Record* 23 (September 1943): 1. NYCBT, *Transit Record* 27 (September 1947): 1. NYCBT, *Transit Record* 28 (September 1948): 1. All figures are for the fiscal year.

2. Maurice Zolotow, "Manhattan's Daily Riot," *Saturday Evening Post* 217 (March 10, 1944): 22.

3. Murray Schumach, "What's Wrong with the Subway? What Isn't!" *New York Times Magazine* (March 10, 1946): 14.

4. Quoted in Ibid.: 88.

5. "The Reminiscences of Newbold Morris" (1950): 4, Oral History Collection of Columbia University, Butler Library, Columbia University, New York, NY.

6. Wilfred Owen, *The Metropolitan Transportation Problem,* rev. ed. (Garden City, NY: Anchor Books, 1966): 243–44. *American Public Transit Fact Book 1985* (Washington, D.C.: 1985): 31. U.S. Department of Transportation, *Federal Highway Statistics Summary to 1975* (Washington, D.C.: U.S. Government Printing Office, 1975): 45. NYCBT, *Transit Record* 25 (September 1945):

1. NYCTA, *Transit Record* 40 (September 1960): 1. NYCTA, *Annual Report 1961* (New York: n.p., 1961): 7.

7. Maurice Forge interview, May 9, 1987, LaGuardia and Wagner Archives, LaGuardia Community College/CUNY, Long Island City, NY.

8. Ibid.

9. *National Cyclopaedia of American Biography,* v. 54 (Clifton, NJ: James T. White and Company, 1973): 6–7. *New York Times,* December 16, 1967. Thomas Kessner, *Fiorello H. LaGuardia and the Making of Modern New York* (New York: McGraw-Hill, 1989): 69. Telephone interview with Paul Windels, Jr., December 9, 1991.

10. Committee of Fifteen, *Recommendations to Civic, Labor, Business, Civil Service and Taxpayer Groups for a Fiscal Program for New York* (New York: n.p., 1943): 2.

11. "The Reminiscences of Paul Windels" (1950): 418, Oral History Collection of Columbia University, Butler Library, Columbia University, New York, NY.

12. Committee of Fifteen, *Recommendations to Civic Groups:* 1–2, 10–13. Committee of Fifteen, *Report on Transit Fares* (New York: n.p., 1943): vi. *New York Times,* December 16, 1967. Citizens' Transit Committee, *The Subways of New York* (New York: n.p., 1944): 20–21.

13. Committee of Fifteen, *Recommendations to Civic Groups:* 1–3. Committee of Fifteen, *Report of Transit Fares:* vi.

14. *New York Herald-Tribune,* February 7, 1925; December 23, 1930. *New York Times,* February 20, 1927.

15. *New York Times,* March 10, 1925.

16. *New York Times,* March 6, 1947.

17. Committee of Fifteen, *Recommendations to Civic Groups:* 1–2, 10–13.

18. Committee of Fifteen, *Recommendations Concerning the Finances of New York City* (New York: n.p., 1944): 12.

19. Paul Windels, "Statement," WMCA Business Forum Broadcast No. 37, A New Era for the Straphanger, November 24, 1943, box 10, SMI.

20. "Extracts from Statement made by Stanley M. Isaacs Representing United Neighborhood House and New York City Consumer Council, Before the Board of Estimate on February 10, 1947," Subway Fare folder, box 15, SMI. Stanley M. Isaacs, "Statement for WMCA on Five-Cent Fare," WMCA Business Forum Broadcast No. 37, A New Era for the Straphanger, November 24, 1943, box 10, SMI. Terry S. Ruderman, "Stanley M. Isaacs: The Conscience of New York" (Ph.D. dissertation, City University of New York, 1977): 1–7. *Who's Who in New York (City and State), 1952,* 12th ed. (New York: Lewis Historical Publishing Company, 1952): 578–79. J. M. Flagler, "The Public Be Served: Part 1," *The New Yorker* 35 (December 12, 1959): 24.

21. "The Reminiscences of Stanley M. Isaacs" (1950): 256, Oral History Collection of Columbia University, Butler Library, Columbia University, New York, NY.

22. Ibid.: 256.

23. Robert A. Caro, *The Power Broker: Robert Moses and the Fall of New York* (New York: Vintage Books, 1975): 837–39, 843, 894–919. Triborough

Bridge Authority, *Construction Schedule for Arterial Highways and Parkways* (New York: n.p., 1945): 1–7.

24. Stanley M. Isaacs, "Memorandum—Five-Cent Fare," June 1946, Subway Fare folder, box 14, SMI.

25. Joshua B. Freeman, *In Transit: The Transport Workers Union in New York City* (New York: Oxford University Press, 1989): 288.

26. "Windels Reminiscences": 122–24.

27. NYCBT, *Report Including Analysis of Operations of the New York City Transit System for Five Years Ended June 30, 1945* (New York: n.p., 1945): 25–31. "Windels Reminiscences": 122–24. Freeman, *In Transit:* 227–28.

28. Peter V. Cacchione, "Save the 5¢ Fare," *Coney Island Pilot* 2 (February 1945): 1.

29. Gary M. Fink, ed., *Biographical Dictionary of American Labor Leaders* (Westport, CT: Greenwood Press, 1974): 296–97. L. H. Whittemore, *The Man Who Ran the Subways: The Story of Mike Quill* (New York: Rinehart and Winston, 1968): 3–50. Freeman, *In Transit:* 286–317. "Quill Resigns from N.Y.C.I.O. Council in Policy Split," Federated Press release, March 29, 1948, Michael J. Quill folder, Biographical file, box Q-Rea, Federated Press Papers, Division of Rare Books and Manuscripts, Butler Library, Columbia University, New York, NY.

30. "Mayor O'Dwyer's Speech on the Transit Fare Problem," press release, radio station WNYC, April 20, 1948, Transportation folder, box 969, William O'Dwyer Papers, Municipal Archives, New York City Department of Records and Information Services, New York, NY. Esther L. Moscow to William O'Dwyer, Subway fare—trend mail—supplement memorandum, April 22, 1948, Mayor, Jan.–June 1948 file, box 968, O'Dwyer Papers.

31. John C. Bollens, *Special District Governments in the United States* (Berkeley, CA: University of California Press, 1957): x. James F. Richardson, "To Control the City: The New York Police Department in Historical Perspective," in *Cities in American History,* ed. Kenneth T. Jackson and Stanley K. Schultz (New York: Alfred A. Knopf, 1972): 274–77. Paul Studendski, *The Government of Metropolitan Areas in the United States* (Washington, D.C.: National Municipal League, 1930): 264–65, 274, 279–85.

32. Michael N. Danielson and Jameson W. Doig, *New York: The Politics of Urban Regional Development* (Berkeley, CA: University of California Press, 1982): 40. Erwin Wilkie Bard, *The Port of New York Authority* (New York: Columbia University Press, 1942): 3–45, 122.

33. Danielson and Doig, *New York: The Politics of Urban Regional Development:* 177–94. I am grateful to Jameson W. Doig for generously providing me with information about the Port Authority.

34. Caro, *Power Broker:* 429–34, 475–80, 613–14, 671. *New York Times,* December 16, 1917. *National Cyclopedia of American Biography,* v. 54 (New York: James T. White and Company, 1973): 6–7. Bard, *Port Authority:* 179, 184, 188, 190–93.

35. "Windels Reminiscences": 53.

36. *New York Times,* March 23, 1953.

37. *New York Times,* February 11, 1945; November 28, 1948. *New York World-Telegram,* April 21, 1948.

38. *New York Times,* January 18 and May 8, 1949. Stanley M. Isaacs, "Notes for ADA City Affairs Commission Session," September 12, 1952, Speeches folder, box 21, SMI.

39. A. A. Berle, Jr., to Fiorello H. LaGuardia, Samuel Seabury, and Paul Windels, "Memorandum re. Transit Unification—Securities to Be Exchanged," February 23, 1934, Transit—Seabury-Berle Plan—Hearing and Disposition folder, box 2590, FHL.

40. Boris S. Pushkarev with Jeffrey S. Zupan and Robert S. Cumella, *Urban Rail in America: An Exploration of Criteria for Fixed-Guideway Transit* (Bloomington, IN: Indiana University Press, 1982): table A-2. T. C. Barker and Michael Robbins, *A History of London Transport,* v. 2, *The Twentieth Century to 1970* (London: George Allen & Unwin, 1974): 270–311, 337–58. E. Edon Barry, *Nationalism in British Politics: The Historical Background* (Stanford, CA: Stanford University Press, 1965): 284–303. *Jane's Urban Transport Systems 1984* (London: Jane's Publishing Company, 1984): 246–47, 261–63, 329–31, 337, 353, 377–79, 424–25. This trend continues today. Many of the world's most important metros—principally Seoul, Mexico City, and Washington, D.C., but also Vienna, Rome, and Prague—are situated in national capitals. For the rise of the federal role with the Urban Mass Transit Act of 1966, see Alan Altshuler with James P. Womack and John R. Pucher, *The Urban Transportation System: Politics and Policy Innovation* (Cambridge, MA: MIT Press, 1981): 32–37, 301–4.

41. *New York Times,* February 8, 1952.

42. *New York Times,* February 8, September 21, and October 17, 1952.

43. Thomas E. Dewey to the legislature, March 10, 1953, series 6, box 199, folder 2, Thomas E. Dewey Papers, Department of Rare Books and Special Collections, Rush Rhees Library, University of Rochester, Rochester, NY.

44. *New York Times,* March 11, 1953.

45. Philip Hoffstein to Vincent Impellitteri, April 29, 1953, Board of Transportation, Jan.–May 1953 folder, box 1345, Vincent R. Impellitteri Papers, Municipal Archives, New York City Department of Records and Information Services, New York, NY.

46. F. W. Grumman to Stanley M. Isaacs, April 6, 1953, Fare folder, box 23, SMI.

47. New York State, *Laws of the State of New York,* v. 1 (Albany: n.p., 1953): 746–59. *New York Times,* March 11, 17, and 20, 1953.

48. *New York Times,* March 21, 1953.

49. *New York World-Telegram and Sun,* June 2, 1953. *New York Herald-Tribune,* June 3, 1953. *New York Times,* April 20 and March 23, 1953.

50. New York City Transit Authority, *Agreement of Lease Between the City of New York and the New York City Transit Authority* (New York: n.p., 1953): 2–8. *New York Times,* June 2 and 15, 1953.

51. Robert Moses, George V. McLaughlin, and Charles G. Meyer to William O'Dwyer, October 7, 1949, Board of Transportation, September–October 1949 folder, box 974, O'Dwyer Papers.

52. Kenneth T. Jackson, *Crabgrass Frontier: The Suburbanization of the United States* (New York: Oxford University Press, 1985): 283–305.

53. Stanley M. Isaacs to Wallace Andrews, September 19, 1956, City Coun-

cil, N–O folder, box 28, SMI. An examination of the *New York Times Index* indicates that there was less coverage of New York's mass transit lines from the mid–1950s to the early 1960s than at any other point since 1900, except during the two world wars. The following are the total number of pages used to list the references under the heading "New York City Transit System": 1950—5; 1951—5; 1952—7; 1953—11; 1954—6; 1955—6; 1956—6; 1957—5; 1958—4; 1959—4; 1960—3; 1961—4; 1962—3; 1963—2; 1964—3; 1965—5; 1966—8. Every fraction of a page was counted as a full page. Source: *New York Times Index* for 1950 through 1966.

54. *New York Times,* February 6, 1970; February 15, May 17, and June 30, 1984. "Trouble Underground," *Headlights* 32 (June 1970): 4–8. Telephone interview with David Ross, February 9, 1990. *New York Times,* November 6, 1967; October 3, 1974; September 30, 1985.

Epilogue: The Kitchen Debate

1. *New York Times,* March 12 and 15, April 8 and 10, June 7, July 5 and 22, and September 2, 1959. Michael R. Beschloss, *Mayday: Eisenhower, Khrushchev and the U-2 Affair* (New York: Harper & Row, 1986): 179–82. The American National Exhibition also featured a "typical" city family and a "typical" union family, who, like the Davises, were selected by the Fashion Industries Council, based at New York City's Fashion Institute of Technology. Significantly, the Davises received much more publicity than the other two families and were clearly understood to embody the American archetype. *New York Times,* May 15, 1959.

2. *New York Times,* March 15 and July 25, 1959. William Safire, *Before the Fall: An Insider's View of the Pre-Watergate White House* (Garden City, NY: Doubleday and Company, 1975): 3–7. Nikita S. Khrushchev, *Khrushchev Remembers: The Last Testament,* trans. and ed. by Strobe Talbott (Boston: Little, Brown and Company, 1974): 364–67. Richard M. Nixon, *Six Crises* (Garden City, NY: Doubleday and Company, 1962): 242–63.

3. Beschloss, *Mayday:* 181.

4. *New York Times,* July 26, 1959.

5. Sigurd Grava, "The Metro in Moscow," *Traffic Quarterly* 30 (April 1976): 242–67.

6. *New York Times,* July 26, 1959.

7. Metropolitan Transportation Authority, *No Standing Still: The MTA Capital Program Phase 3, 1992–1996* (New York: Metropolitan Transportation Authority, 1991): 2–5.

INDEX

MAP SOURCES

Map 1: "A Great Rapid Transit System for a Great City," *Scientific American* 103 (December 17, 1910): 478.

Map 2: "A Great Rapid Transit System for a Great City," *Scientific American* 103 (December 17, 1910): 478.

Map 3: New York State Transit Commission, *New Subways: Proposed New Additions to Rapid Transit System to Cost $218,000,000* (New York: n.p., 1922): 17.

Map 4: Derived from Joseph Cunningham and Leonard O. DeHart, *A History of the New York City Subway System,* v. 3, *The Independent System and City Ownership* (New York: n.p., 1977): 24.

Map 5: Regional Plan of New York and Its Environs, *Regional Survey of New York and Its Environs,* v. 4, *Transit and Transportation* (New York: Regional Plan of New York and Its Environs, 1928): 41.